Medical-Device Usability

This book simplifies the process regarding the application of usability to medical devices for new developers by providing a thought-out walkthrough of what is involved and how the expectations from the essential standards affect their work and why. Based on the authors' experiences, the information in the book provides other more seasoned experts with a wealth of insight and nuggets of new perspectives on addressing the usability of medical devices, particularly medical software.

The discussion in the book is rooted in recent real-world experiences from both the manufacturer and assessment body sides—the authors provide a novel high-level "roadmap" to using usability engineering to guide development activities from initial conceptual design to research, development, and ultimately release of a medical device. They build toward this map throughout the book and ultimately present the full model. This roadmap is the result of the authors' experiences of taking a research project and launching both a medical device and a medical device business after obtaining all the necessary regulatory approvals.

The book helps product developers to understand the usability requirements around their own work, and to also better understand their users and to present the expected usability information in a way that is readily understood by the auditors. Managers of such work will also greatly benefit from the discussion of the required deliverables and the goals of the usability activities.

Finally, the authors ease the divide between manufacturers and healthcare professionals using medical devices. By simultaneously showing the requirements for the manufacturer's development process and also discussing the realities of work on the healthcare side, the authors facilitate collaboration.

Medical-Device Usability

Human Factors for Medical Device Development

Ilkka Juuso and Ilpo Pöyhönen

Routledge
Taylor & Francis Group
A PRODUCTIVITY PRESS BOOK

Designed cover image: Web Large Image (Public)

First published 2026
by Routledge
605 Third Avenue, New York, NY 10158

and by Routledge
4 Park Square, Milton Park, Abingdon, Oxon, OX14 4RN

Routledge is an imprint of the Taylor & Francis Group, an informa business

ISBN: 9781041085065 (hbk)
ISBN: 9781041085058 (pbk)
ISBN: 9781003645702 (ebk)

DOI: 10.4324/9781003645702

Typeset in Garamond
by Newgen Publishing UK

Dedicated to the tigers in the zoo, in the wild, and on the bus with us to making better medical devices

Contents

Contents

Figures

Preface

The usability of medical devices is a topic that is much too easy to either take for granted as something just anyone can do, or alternatively brush under the carpet as something you surely won't have to do, just as swiftly as it is brought up. Being bullish on the topic or burying your head in the sand like an ostrich will only get you so far though. Regardless of where you see your rightful place in this animal kingdom you would do well to take a fearless, opportunistic approach to usability engineering as a methodology that holds the power of showing you whether you well and truly understand your users and how they interact with your devices – and perhaps even making those devices better in the process. Your fearlessness and opportunism will be put in check by your risk management and clinical evidence activities soon enough, and your quality management activities won't let you release a product that does not hold water. That is not to say that usability engineering could be equated to a kindergarten activity or a pricey luxury. Far from it. The point here is that there is a lot of benefit to be gained from both formative and summative usability work which you won't reap if you take a rigid and almost preordained approach to usability engineering or human factors analysis.

Usability engineering should never be a checkbox activity without any effects, and it is not an omnipotent trump card that lets you release just any device you have been wanting to put out there and claim that it is safe. Usability engineering is not about putting crash test dummies in a vehicle and waiting to see how badly they are injured. It is not about stacking your eggs on the backset of a T-model Ford and driving around your church five miles an hour to see if all the eggs could theoretically survive. Instead, usability engineering should be a practical method for communicating with your users, learning what you should do to empower them to use your devices correctly, and then after all that is done feel confident enough that you have made the most out of all the lemons you've been handed.

Usability engineering shouldn't just be an isolated activity within your organization either – too often we have seen shortcomings in executing on the many good observations made during usability evaluations as these have failed to become requirements or see final implementation. Usability engineering is not a checkbox to tick, but it is often a major part of gaining the necessary confidence to tick the final checkbox for releasing your product to the world.

In this book, we will dive deep into usability engineering with the international IEC 62366-1 standard on the application of usability to medical devices as our principal guide. We will interrogate the standard mercilessly to arrive at the true meanings behind each of its requirements and present these to you against the practical framework for medical device development that we had introduced in our two previous books: "*Developing an ISO 13485-Certified Quality Management System – An Implementation Guide for the Medical-Device Industry*" (Routledge 2022) and "*Medical-Grade Software Development – How to Build Medical-Device Products that Meet the Requirements of IEC 62304 and ISO 13485*" (Routledge 2024). We don't pack any brass knuckles in our travel kit, and we don't have an axe to grind with the standard, so we might as well be wearing silk gloves here, but we promise we won't stop here before the requirements make sense and fall into line with everything expected of modern medical device development and quality management.

Through the sections of this book, we will impart with advice we feel anyone embarking on a usability engineering project should know, starting with what to know before starting off on such a trek in Section 1. This section is written to shoot down easy misunderstandings – such as thinking that medical-device usability would not be more about safety than it is about ease of use – and to get us on the same page for the sections to come. Section 2 takes us on a tour of the fundamental IEC 62366-1 standard. Here we don't just point out the clauses to you like the lions and elephants on a safari, but we put each clause through a real physical examination and give you the vital statistics on each of them. This section packs a punch and represents the highest hill we need to climb on this expedition. Section 3 then continues with a practical overview of the various usability engineering tools you may want to employ in meeting requirements. The strongly related Section 4 then represents the travel journal we all need to write of our experiences with the users, i.e., it presents our model for the expected usability engineering file. Here we will take a comprehensive look at the expected documentation of usability engineering activities, and how you might fashion this into your overall medical-device records. Section 5 allows us to jump the fence of the zoo and step into the shoes of the doctors, nurses, other medical practitioners and even the patients themselves who are the users of our medical devices. Section 6 empowered by an understanding of the user then looks at how we might address the benefits associated with our device. In Section 7, we address the lines of practical communication and transparency we might want to keep up across the compound fence, both now and over the long term in order keep an ear to the ground on how our devices are doing in the real world and to keep the user abreast of important changes to the devices. Section 8 looks at the present and future of listening to the user. Section 9 looks at usability against the backdrop of our full quality management activities and investigates how usability engineering might factor into these activities. Section 10 goes even further by attempting to sketch out a paint-by-numbers approach to how usability can inform all our medical-device development activities from concept to a finished product. This model should prove particularly useful for startups, but not only startups. Section 11 discusses foreseeable future challenges to usability engineering from the evolving use of medical devices to the use of artificial intelligence. Section 12 walks us through conformity assessments for usability. Section 13 then provides a few departing lessons on the topics we have covered in this book. Finally, Section 14 looks back at the zoo we have created and hopefully is prouder of the result than John Hammond flying off his beloved theme park at the end of Jurassic Park. The section aims to put a pretty ribbon on our deliberations while pointing out any hairline fractures in the foundations we have poured.

None of the animal action on this expedition will be monitored by the American Humane Association, unlike what you may have seen at the end of Hollywood films. Here we take a Descartian approach to dismantling the standard clause by clause, and after seeing what it is made of, figure out how it changes what we know about developing medical devices. The ultimate goal is to find out how we might negotiate all this together under the established ISO 13485 umbrella of quality management for medical devices in an inherently lean, practical, and safe way.

We hope you will find this expedition insightful, entertaining, and to the jugular. Please let us know of your experience, feedback, and any comments this book may prompt. Welcome to the domesticized jungle!

Acknowledgements

The book you hold in your hands is the third part of our trilogy of books on key medical-device standards. Putting pen to paper on the first book, on ISO 13485 quality management, we never thought we'd end up here, but the journey has proven to be both educationally rewarding and quite unexpectedly also deeply addictive. This triathlon of a dive has given us the license to plunge headfirst into the core of medical device development to distill the requirements involved in the quality management, risk management, and software development activities around all medical devices into a practical set of rules to work by.

In movies the third installments rarely are worth the wait. Think of Terminator 3, the comparatively underwhelming Godfather Part III, and the profoundly unsatisfactory Matrix Revolutions. Occasionally, though, we do get a delicious exception to this rule in the form of films such as Back to the Future Part III, Die Hard 3, Alien 3, or Indiana Jones and the Last Crusade. Even Naked Gun 33 1/3 was enjoyable in its own right, whether properly chilled or not. Such an exception is what we are also going for here. Here we hope to go back to basics, solve the riddles around understanding the user, and meticulously sketch out the notebook you can tuck in your pocket as you venture out to meet your potential users. In other words, we hope to pull you back in for another dive and convince you that usability engineering may hold keys to more questions than just what a pretty interface should look like.

For us this book is not an afterthought to the two previous two installments, but rather it's a book we have wanted to write from the very start. In writing our second book on medical-software development we often delved into usability-related discussions, which we occasionally commented on in that book, but more often than not wrote down on digital post-it notes to inform our discussion here.

Much has happened since the second part of this trilogy came out. For one thing, we have now both parked our wagons on the certification side of the medical-device business. In doing so, we have considered the full 360 degrees of the equation to making medical devices, from concept to product, and from both the manufacturer's point of view and that of conformity assessment bodies. As always, there is no single correct answer to any complex issue, but we are excited about what we have to say over the next few hundred pages of this book.

All throughout the project we have been delighted to lean on further experts in medical devices, clinical medicine, and usability engineering across the globe. Most of all, we wish to thank Dr. Risto P. Roine, MD, Dr. Antti Viitala, MD, and Virpi Lahdenmäki for their invaluable feedback on our manuscript and usability engineering in the domain of medical devices. Similarly, we want to thank Michael Sinocchi who as our publishing editor at Routledge in New York is a valued co-conspirator on this trek to greater medical-device insight. Finally, we must thank our families for their love and understanding in letting us invest the time required by this book project. Thank you all!

About the Authors

Dr. Ilkka Juuso, DSc, has 25-plus years of experience working on multidisciplinary R&D projects in both industry and academia. He is one of the founders of the medical-device startup Cerenion, a former senior advisor and principal regulatory engineer at the medical-device quality consultancy Kasve, and a post-doctoral researcher with the University of Oulu in Finland. Today he works as a lead auditor at the European notified body SGS Fimko (NB 0598) on conformity assessments according to the EU 2017/745 ("MDR") and international standards.

His main interests are international regulatory affairs, standardization, and healthcare business development. He has successfully led the development of an ISO 13485, ISO 14971, and IEC 62304 compliant Quality Management System (QMS) from the ground up, its subsequent day-to-day operation, and certification by a notified body. He has had a key role in the launch of a CE-marked Class IIb medical device based on artificial intelligence.

He has repeatedly served as a committee member and the head of the Finnish national delegation in key committees of the International Organization for Standardization (ISO). He is a member of the management team for CEN TC 251 on health informatics, and secretary for its WGII on technology and applications. He is a member of the CEN/CENELEC Strategic Advisory Group on Healthcare and a member of the RAPS Nordic LNG leadership team. He is an active member of the Finnish Standards Association (SFS) and the industry forum Healthtech Finland, and a frequent visitor to International Medical Device Regulators (IMDRF) meetings. He is also the author of the books *Developing an ISO 13485-Certified Quality Management System* (Routledge 2022) and *Medical-Grade Software Development* (Routledge 2024) – and approximately 3,500 film reviews, which probably explains some of the film references in this book.

Ilpo Pöyhönen has 30-plus years of experience working on medical device research, development, testing, and safety and performance evaluation including in the context of an accredited certification body. During that time, he has performed approximately 200 software evaluations according to IEC 60601-1-4, IEC 60601-1 Clause 14, IEC 62304, and IEC 82304-1. His particular areas of interest in this work have been the role of programmable database systems and the development of test equipment for diverse needs. The work has taken him across the globe and even to the edge of space.

Today his main interests are international regulatory affairs, standardization, and the intelligent control of medical device design processes to continuously meet the requirements imposed by, for example, cyber security, usability engineering, risk management, and agile development models. The use of emerging technologies, such as artificial intelligence, also holds special appeal to him.

He has been active in research initiatives that have, for example, examined the software development documentation required in a regulated environment, the impact of risk management, the performance of risk analysis itself as part of the software development life cycle, and the reliability factors involved in the supply of complex software systems.

He is a long-time committee member of SFS/SR301 on healthcare IT and a sought-after lecturer on topics such as medical device software, risk management, usability, mHealth apps, and cloud services in the context of medical devices. He is also the co-author of the book *Medical-Grade Software Development* (Routledge 2024).

Chapter 1

What to Know Before Getting Started

We approached writing this book with an aim of pointing out what truly matters without too much fanfare, and all the while making the biggest possible bang with each single sentence we put in front of you. If the language is occasionally colorful, or infused with a film reference or two, we hope that these too will delight more than detract and even add to your understanding via some auspicious neurological association connecting the dots and illuminating concepts in new ways. This is how we approach the whole book, but perhaps nowhere are these goals more visible than here in this first section as we hope to sum up the biggest takeaways and concepts from usability engineering. This section discusses the top 25 lessons we feel everyone should know before diving into usability engineering in the context of medical devices. The lessons here are written from the point of view of thwarting off some easy misunderstandings on the topic and getting us all on a safe footing for the discussion ahead.

Here we appreciate the utility of wearing a gorilla suit in some zoological simulation but do away with any pretense of the suit making us experts on the real-life behavior of the said quadrupeds. In cinematic terms, Eddie Murphy and Dan Akroyd may have found a good use for such suits in "Trading Places", but to really trade places with gorillas we would probably be better off following Sigourney Weaver's lead in "Gorillas in the Mist" and observe our real users in their natural habitat. The objective here is to learn how to step into the shoes of users – even if that user can sometimes be

DOI: 10.4324/9781003645702-1

King Kong and wear armored tanks for shoes and pick his ears with palm trees – before we are faced with things having run amok and the Planet of the Apes calling the shots.

Now that we have quickly exhausted our quota of furry film references here, let's get into the lessons themselves. The following are the top lessons we would like you to consider in order to get off on the right footing to understanding the zoo of usability engineering. We call it a zoo in the most affectionate of terms, but also because we feel this is an apt analogy for aiming to understand the jungle out there via cleanly delimited simulations and the occasional necessarily constrained safari. We try to leave all monkey business to the Marx brothers, though.

At the end of the book, in Section 13, we'll present a few more lessons relevant to running the zoo over any extended periods of time. For now, though, let's focus on brushing aside the foliage and poking our heads to the world of usability engineering.

1.1 Medical-device usability is almost exclusively about safe use

The first lesson to know regarding usability engineering in the context of medical devices is that it really is about enabling safe use instead of looking good while doing so. Think of adult diapers instead of Victoria's Secret. Think safe, unoriginal movie sequels with name actors and a story you have seen and sort of approved at least a dozen times before. Think of a doomsday device with an unbreakable lid as depicted in Figure 1.1. *There is more to medical-device usability but safety is always the first goal.*

Adult diapers are perhaps the most uninspiring example we could think of here. No matter how cool you make the package, how snazzy you make the advertising campaign, and how off-topic you make the user training this is not a topic that captures most people's imaginations. Diapers usually aren't even medical devices unless they contain some fancy features such as sensors and artificial intelligence for a medical purpose. Yet, diapers are the perfect example to set the baseline for our mission here. There is much more to usability, thankfully, but to discuss all that we need in terms of understanding a product, its users, use contexts, and uses, we could make diapers the red-thread example throughout this book. We won't though.

Figure 1.1 A detonation button behind unbreakable glass might be safe but unusable.

In the following sections of this book, we will dive deep into usability, and how to ensure the safe usability of medical devices. We will also expand on this to look at how good usability will be beneficial to medical devices as supporting the safe use of those devices also when no harm directly hangs in the balance. We will even occasionally briefly venture out into the broader, much-trodden domain of general usability beyond any discussion of safety or medical devices but make no mistake: safety is the primary goal any time usability and medical devices come into contact with each other. Figure 1.2 illustrates the overlap between what is in general considered as usability and what is considered as safety in the highly regulated context of medical devices.

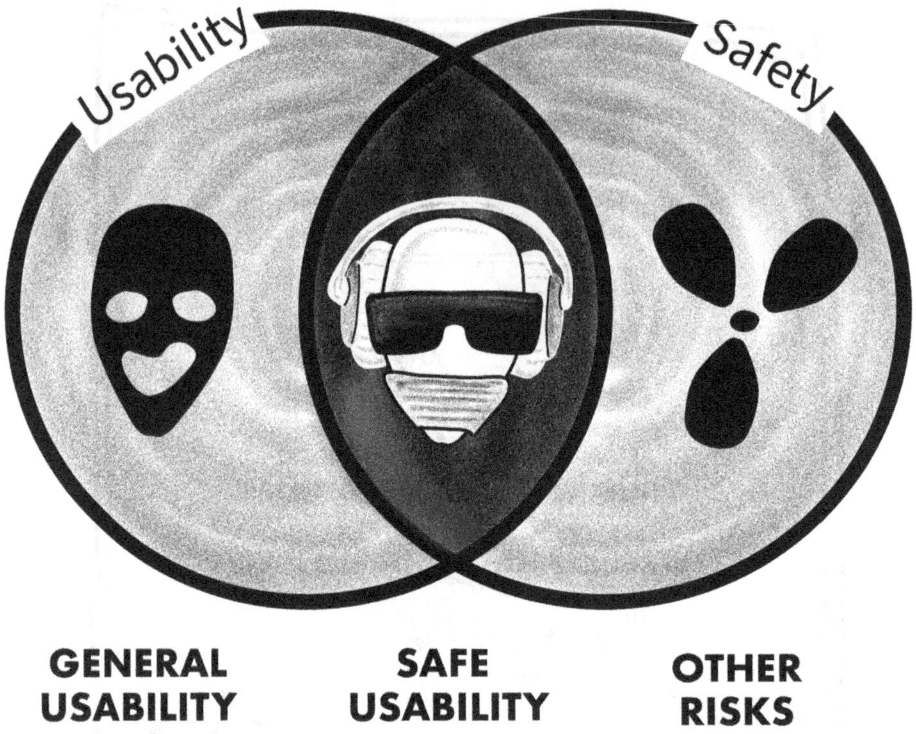

Figure 1.2 Usability, safety, and safe usability.

The figure shows how the area of joint interest for usability engineering and the development of medical devices is represented by the application of usability engineering to developing devices that are safe to use ('safe usability' in the figure). The international standard on the application of usability engineering to medical devices, IEC 62366-1, is at the very center of this. The standard will be a big part of the remaining discussion within these covers in both showing us what we need to be working towards and helping us understand why. We will distill the requirements of the standard to their cores and explain their underlying aims. You may also have other device-specific standards, such as those belonging to the IEC 60601-1 family of standards, to help you consider all the relevant aspects of safety around the type of medical device you are developing. These standards may point out safety-critical features and uses for you to consider in your overall usability engineering activities (see Section 1.5).

Not all risks related to a medical device arise out of usability issues, of course. To address such other risks, we will occasionally refer to the international standard of risk management for medical devices, ISO 14971.

This is also referenced in several places by IEC 62366-1 itself and is an essential part of the whole picture of addressing the use of medical devices and the possible harms involved therein. Here too the emphasis is squarely on safety. Risk management, though, is not the main topic for this book. For more discussion of the risk management standard please refer back to Juuso 2022 and, for a software-focused discussion, Juuso & Pöyhönen 2024.

Finally, we will also discuss general usability engineering as it aims at not just safety but some broader goal of good usability (see Section 2.6). Usability is a topic that goes beyond just medical devices, and one that may occasionally be considered art instead of science, but it certainly benefits from more stakeholder viewpoints than just the engineers who made the device in question. We say that as engineers ourselves, but as what we hope would qualify as fairly humane engineers.

In our discussion of usability engineering, we also make use of a technical report published as IEC 62366-2 and thus related to the standard itself, IEC 62366-1. The technical report is not a required standard; its adoption is very much optional, but it nonetheless serves as a good indication of how the medical-device world looks at usability more generally and what tools it sees as part of the work in usability engineering. The technical report, or guidance as we shall mostly refer to it, discusses a variety of tools that can be used in usability engineering for both safety and non-safety goals. The guidance does not add any new requirements to usability engineering, although it does occasionally have a hard time holding back in its place as a purely supportive, informative piece of literature (see, e.g., usability engineering report in Section 4.1.8). In Figure 1.2, the IEC 62366-2 guidance is most accurately placed in the overlapping region between general usability and the medical-device usability standard as it attempts to give helpful advice on meeting the requirements, but is not a required part of the picture. It is also worth pointing out that the TR is currently in the process of being rewritten as a TS, a technical specification, but it still won't be a part of the hard requirements required for usability engineering.

In this book, we'll first look at usability engineering through the safety requirements placed on it in the context of medical devices (Section 2), then discuss the recommended tools in more detail (Section 3), and finally go on to present our model of documentation for use in the usability engineering file based on both the requirements and the recommendations (Section 4). For now, though, it is enough to remember that when it comes to the usability of medical devices a call to safety is the loudest call sounded over the jungle, not some seductive mating call dreamt up by an advertising department.

1.2 Stepping into the shoes of the user

You don't know a man until you have walked a mile in his shoes, goes the old saying. But what if the user wears high heels, flipflops, or has a prosthetic leg and dislikes honest doctors as much as that killer in the Harrison Ford film The Fugitive*? Stepping into the user's shoes is no small feat, as* Figure 1.3 *tries to illustrate.*

Good usability is a deceptively familiar quality that is easy to dismiss as something everyone intuitively understands once they see it. It is infinitely harder to define in advance, though, and harder still to define in some practical, instructive manner. This is like taking it for granted that

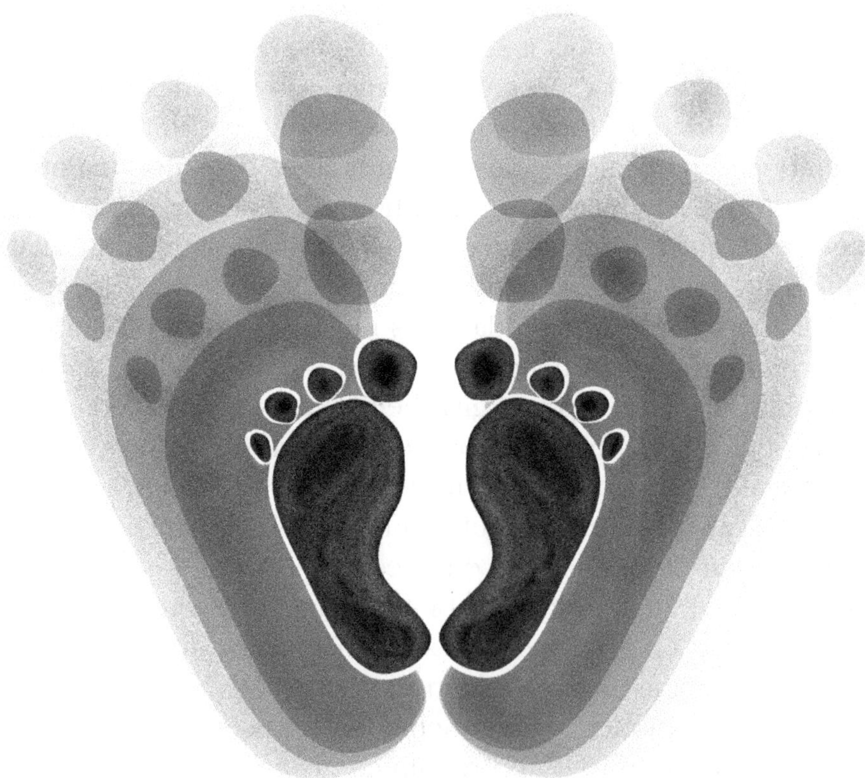

Figure 1.3 Stepping into the shoes of another person is famously difficult.

everyone around the world can speak English but forgetting that a very small percentage of those people speak the language as their native tongue, let alone master it – and the two populations here may not even completely overlap. After all, your nephew may have been speaking the language since the age of two, but is that reason enough for you to trust your corporate communications to him? The situation is much the same for engineering good usability.

The list of usability models and frameworks is at least as long as the roster of action heroes in the continuing Expendables film franchise. Quite possibly all these models could be used to improve medical-device usability, and just as many could be used to wreck any semblance of good usability just the same. In medical devices, the yardstick for measuring all those models, as well as the set of required considerations and documented information, is given by the IEC 62366-1 standard on the application of usability engineering to medical devices. It thus makes sense to map any usability engineering workflow you are accustomed to using to the requirements made by this standard, or to build your usability engineering processes based on the standard from the start.

Deep in the belly of the standard, in its nonbinding annex A3.16, the IEC 62366-1 standard succinctly defines usability as something that enables the user to:

- Perceive information
- Understand information
- Act on information
- Interact with the medical device (to achieve specified goals)

The above listing is not intended as a definition for the term, but it inadvertently hits the nail on its head. You know, like Mel Gibson does in Danny Glover's attic with a nail gun when encountering one of the bad guys in Lethal Weapon 2 – thereby also demonstrating that great usability doesn't lean on manuals or prior training but has the device ready to use and save (some) lives in just seconds. In any case, the four characteristics listed here are at the core of developing intuitive, useful, and safe medical devices with a purpose. A device that does not show information is a black box. A device that does not support its user in understanding the information displayed is a cruel joke. A device that does not allow action to be taken (or withheld) based on its contribution is a useless toy. A device that does not allow itself to be operated is not a tool at all. There are many lenses and competing coordinate systems to view the topic of usability through, but as such models go, the one presented here, by the standard, is a very practical one. For more discussion of perception, cognition, and action see Section 3.5.

IEC 62366-1 defines usability in Clause 3.16 as a characteristic of the user interface that somehow facilitates use and enables effectiveness, efficiency, and user satisfaction in the intended use environment. It is noteworthy that the standard here invokes the intended use environment instead of the intended use of the device, and thus broadens the scope of study from merely the intended use of the device to all use of the device in its intended environment. In practice, the intended use of the device is at the core of any investigation into the usability of that device, while the definition of what other use might be considered reasonably foreseeable is open to much debate. Similarly, effectiveness, efficiency, and user satisfaction primarily matter through their effect on the safety of the device, as observed by the standard in a non-binding note to this same clause.

Interestingly, as much as the IEC 62366-1 standard takes safety at heart, 'safety' is not actually defined as a concept by the standard. It is, however, defined as freedom from unacceptable harm by the ISO 14971 risk management standard and by the software-focused IEC 62304 standard. See Juuso & Pöyhönen 2024 for more discussion of the concept. This is the definition we are obliged to use here too.

The standard also takes a stab at defining the terms 'usability engineering' and 'human factors engineering' in its Clause 3.17. This pair is given meaning as the application of knowledge about human behavior (including abilities and limitations) to the design of medical devices, systems, and tasks. A note given in annex A3.17 explains that the standard sees the two terms as synonymous but appreciates that others may differentiate between the two. Usability engineering may have a slight European flavor to it, while human factors are what the US FDA speaks mostly of. Accordingly, the EU 2017/745 ("MDR") talks of 'usability' (2 times), 'use errors' (2 times), and 'ergonomic features' (3 times), but it does not mention 'human factors' at all. Conversely, the FDA guidance on "Applying Human factors and Usability Engineering to Medical Devices" naturally speaks of 'human factors' (125 times) and 'use errors' (101 times), but less so about 'usability engineering' (11 times) or 'usability' (25 times). It is, of course, profoundly unfair to compare a 175-page piece of general regulation and a 49-page topic-specific guidance document, but the numbers illustrate the point we made.

The standard also takes the rather sobering stance that the well-intentioned application of common sense is not sufficient for good usability. Instead, a systematic usability engineering approach to developing medical-device usability should be adopted from the beginning of the design process to its end. Some, including us, might argue that an element of usability engineering should be a part of all the activities that take place over the entire lifecycle of the device, to the extent that is beneficial, of course. We also take a slightly heretical view to a monolithic summative evaluation when it comes time to revise the work: while a single summative evaluation may work fine as a model for a single instance of a product development project it may not tie into the later change control expected of that device in any optimal way. We fear that redoing a summative evaluation in its entirety will be a task too often postponed when it would be more beneficial to consider usability, as appropriate, whenever making changes to a product. There is great value in a summative evaluation, but unless the changes are also managed in a practical way the ground under that evaluation will change unnoticed.

1.3 Users, user groups, and the right N

In movies you can often spot the heroes and the bad guys just by how they are portrayed leading up to the big life-defining moment the plot will throw their way. As Figure 1.4 *depicts, heroes classically wear white and baddies wear black, at least in old westerns, but not just there. Today the protagonists are perhaps more diverse overall, but after reviewing 3,500-plus films over the past few decades and making it a hobby to predict character fates in films, we have noticed that more often than not these predictions turn out to be somewhat prophetic. There is utility to such stereotypes, even if they don't give you the full picture. In usability engineering stereotypes are replaced by personas and user profiles.*

This may seem self-evident, but IEC 62366-1 Clause 3.24 does provide a definition for a user. Here a 'user' is defined as a person interacting with the medical device. Interaction is to be interpreted as widely as is practical or necessary in the context of a specific medical device. Pressing the buttons of a device to achieve a goal is, of course, interaction but so is accidentally pressing buttons while cleaning that device, and potentially also causing buttons to be pressed by dropping cardboard boxes of the devices on the floor. It may seem like a bridge too far to start considering

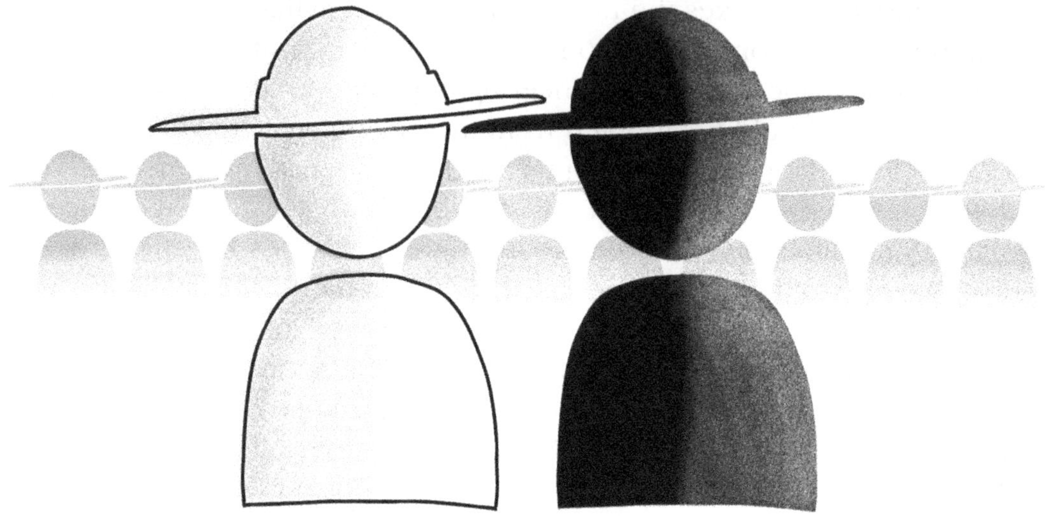

Figure 1.4 Cowboys in white and those in black have quite different scenarios in mind.

accidental use as part of the hazards and harms surrounding your devices, but at least in some cases this can be much more relevant than it might seem at first. A surprising example is given by Neil Armstrong who during the 1969 Apollo 11 mission accidentally backed into a control panel in the lunar module and broke off the circuit breaker for the module's engine. If not for some smart thinking by Buzz Aldrin and the use of a contraband ballpoint pen, the duo might not have been able to return from the moon at all. Resisting butt presses was obviously not a part of the design of the switches, but perhaps some protection against accidental pressures should have been. Don't start looking for butterflies on the far side of the world here but do consider as many of the hazards and harms that may befall your device and its users as is appropriate in its context.

As you would expect, there can be several different types of users including clinicians, patients, and service personnel. In part, the relevant user groups may depend on the uses and use environments you end up supporting for the device – thus the process of figuring out the use, users, and use environment will most likely be iterative in nature (see Section 10). In addition to the intended use of the device you may also want to consider, for example, foreseeable misuse and the transport, storage, installation, operation (if that is not the same as the intended use in some way), maintenance, repair, and disposal of your medical device. Don't go overboard here and get lost in the wilderness of remote possibilities but instead cast your nets as wide as is fruitful and then reel in your catch as is relevant. It will be up to you to connect the dots between the intended and the tested in terms of use, users, and use environments – and the place to provide the rationale for why these pairings are appropriate is in your user interface evaluation plans (see Section 2.3.7).

Users can be assigned to 'user groups' (Clause 3.25) based on factors that are likely to influence their interactions with the medical device. Note that prior to the 2020 amendment to the standard this definition for user groups was worded with a focus on the usability of the device itself, although the meaning was undoubtedly the same. The old definition also gave examples of possible influencing factors such as age, culture, expertise, occupational skills, job requirements, and expected working conditions. Only the first three of these are retained by the new definition. The influencing factors can be further developed into 'user profiles' (Clause 3.29) summarizing such mental, physical, demographic, or otherwise relevant properties of an addressed user group which are thought to possibly have an impact on design decisions. Examples here might include user knowledge, skills, abilities, and limitations (e.g., visual acuity, reduced hearing, limited dexterity, memory issues). Prior to the 2020 amendment the definition was worded slightly differently and focused on only the intended user group whereas now it is explicit that any user group may be addressed.

A persistent question regarding usability studies is how many users should be involved in the tests? This is really an echo of the bigger question of the appropriate N in your clinical evidence, i.e., what number of subjects is enough for clinical studies to be convincing? The answer here, as in clinical studies, is that it depends. For usability studies the figures generally floated land at around 15–20 subjects per user group for the summative evaluations. For formative evaluations the counts are entirely up to the manufacturer. Reportedly 15 test subjects for a summative evaluation was a request from the US FDA when the standard was being drafted while other stakeholders would perhaps have been happy with a lower figure. Subsequently, IEC 62366-1 does not specify a user group size here, but the nonbinding guidance does attempt to partially answer the question (see Section 3.13 for discussion). We would suggest that the appropriateness of your test group, the match between who you tested with and who you aim to have as your actual users, may matter more than, for example, group size or in what country or geographic region you tested in. This all depends on the specifics surrounding your product and the usability testing you intend to perform, of course. Depending on your environment, reimbursement too may factor in all of this and may have stricter expectations than just regulatory approvals.

The assumption is that there is no overlap between the users assigned to each user group in an evaluation but depending on what your user groups are this might not have to be the case. Depending on the characteristics of your user groups and the tasks expected of them you might, for example, have 10 unique users in two different groups, and thus only 20 users in your overall studies. The focus here is on your final, summative evaluations (see Section 1.13) and no such firm recommendations exist for the exploratory formative evaluations (see Section 1.13) – although 5 users is often quoted as the minimum you might generally be happy with yourself. Recruiting fewer test subjects may seem like it will make your work easier but remember that you may then also be more stuck with any single observation you get. Recruiting more test subjects may make your study more convincing, and your results less beholden to any single respondent, but in the end, it is mostly about how representative the users are that you recruited. Don't think of test subject recruitment as jury selection, though, but instead as roping in appropriate guineapigs. For your device to fly it may thus be more important to recruit the right stuff instead of just the right N – as indeed was the case also when NASA needed to recruit seven astronauts for the Mercury space flight, if you allow the obscure film reference.

Developing various user and user group profiles is an important activity to get right. Defining the characteristics of these profiles in a way that is appropriate to your intended use without unnecessarily splitting up groups (and thereby leading to expectations of greater test populations) matters. At the same time, it may be necessary to observe both similarities and differences between the groups (e.g., in terms of age), and even allow the assignment of the same users into multiple groups in some transparent way. Uncontrolled or invisible overlap between the groups is to be avoided like the plague, but some flexibility may be beneficial to addressing all the user groups of interest to you – particularly as the standard half-expects you to address not just the intended use of your product but all its use in its intended environment. Double-dipping, or assigning the same users to several user groups, may not be a wise choice for your definitive summative evaluations, but it may be necessary for your formative evaluations as you explore different potential user populations and uses.

Defining profiles for collateral groups of users may not be worth the effort, and may actually detract from your focus on the more essential user groups for your device. For example, a laboratory cleaning technician may come into proximity of your device every night but addressing them as a specific user group may not be worth your while. Similarly, it apparently has happened that a husband accidentally pulled the plug on his wife's ventilator when looking to plug in the heater for his car. For these and other reasons you should think carefully about the context your device goes into. Where you draw the line between astronauts and armstrongs is, of course, up to you, but we would always focus on what truly matters and only expand from there as our resources allowed and the particulars of our device required.

1.4 User interface

The interface for the self-destruct mechanism onboard the Nostromo that Ripley is forced to work on at the end of Alien *is perhaps the worst example of an interface in the history of films. That interface wasn't a medical device, but it could have gotten her killed and it did get her pregnant without her consent, even if it did also blow up the ship. Being forced to go through a whole physical exercise routine like that to blow up a spaceship seems silly, but it is, of course, quite preferable to something else where you might accidentally detonate everything when only trying to refresh your social media feed.* Figure 1.5 *illustrates an interface trying to fend off users.*

IEC 62366-1 Clause 3.26 gets somewhat obvious in defining a 'user interface' as the means by which a user and the medical device interact. Nonbinding notes to the clause here do, however, position accompanying documentation (see Section 3) as part of the device and its interface, pull in all interaction elements of the device (including physical, visual, auditory, tactile, and non-software aspects), and suggest that a whole system of medical devices may be addressed via single user interface if appropriate. It should also be pointed out that occasionally the user interface may even have multiple modes or different faces to different users. The first interface we usually remember to discuss is the one used by the medical staff or the patient themselves, but this may or may not be the same interface other user groups such as hospital maintenance staff use. It may thus occur that

Figure 1.5 If your UI resembles an obstacle course you may only get the most devoted of users.

some software update or a flick of a switch in the service menu changes the defaults for some setting (e.g., an alarm threshold) without making it obvious that something will change for the actual user as well. All these different faces of the interface are technically covered by the standard, but their possible existence is not explicitly brought up by the standard.

The message here is to beware of relying on context sensitivity in cases where contexts may differ in, for example, the form of the graphical user interface or the mindset of the person somehow interacting with that facet of the software. We have probably all seen badly localized translations of user interfaces and manuals – one of our personal favorites is the user manual for an old-fashioned rear projection television set, the cover of which in one language read "Throwing TV". The reason for the mistake was an overly simplistic approach to translations where a manufacturer thought a word would always be translated one way regardless of its context, and its translation could thus be used interchangeably anywhere where the original word appeared. Not so. The lesson here is that in addition to differences in language the users may also have different expectations and frames of reference for interpreting any given detail in the user interface; for example a doctor and a maintenance technician might understand the significance of a setting differently.

As the standard strictly defines a user as a person, the standard is not readily applicable to application programming interfaces (API) and other interfaces intended for use between machines. This is a topic occasionally debated during audits of machine-to-machine interfaces – particularly when humans are the ones who first must implement those interfaces before any exchanges can occur – but it is something which the standard is oblivious to. Thus, user interfaces are the interfaces to using the medical device, graphical or otherwise, but according to IEC 62366-1 users are always people. In other words, androids may dream of electric sheep, but they will not get treated for sleep apnea.

We are not suggesting that all APIs should be subjected to usability engineering activities, but in some cases this might be beneficial. Here we are reminded of an electronic health record system where a certain measurement value and the price of that measurement were close enough in appearance so that the price was accidentally chosen for display when the actual measurement value was intended to be shown. The mistake here was made when programming the software, or more specifically, when the programmer was making use of an API providing access to the underlying data. The mistake could have perhaps been prevented by coupling the value with a unit (dollars versus mmol/l), but it could perhaps also have been caught via a usability-instructed review of the API when it was being constructed. Usability may not matter for computers using our devices, but anytime humans are in the loop, be that in using the device or in setting up the use of an API provided by our device, usability engineering may provide valuable tools for us to use.

Following from the above discussion, a 'user interface specification' (IEC 62366-1 Clause 3.28) is the body of specifications that collectively and comprehensively describe the user interface of a medical device ahead of time. The term used here is "prospectively" indicating that the specification is to be developed before the user interface is created, but also at the same time leaving debatable room for tweaks to occur before or during implementation. Annex A3.28 of the standard then builds a veritable apartment complex in this vacant space by stating that the user interface specification should be developed early on and in a way allowing the engineering team to go about implementing the interface – all the while bracing for changes to be made via an iterative approach.

The standard says nothing about denoting the certainty affixed to any specific properties of the evolving specification but somehow flagging the properties that have already been verified and locked, in contrast to other properties which may have yet to be verified or narrowed down to their final characteristics, may be a wise step to take here if an iterative approach is your aim. We don't mean to suggest that a draft specification would somehow suffice at the end, only to observe that very few things are born final and that keeping track of the maturity of your evolving work may be helpful to you during the development work. A system like this may be your best protection against having to tear down both sides of a frontier city street set when the width of that street was later found to be less than perfect for filming, as it happened during the troubled filming of Heaven's Gate. See Section 2.3.6 for further discussion of the user interface specification.

Finally, user interface evaluation (IEC 62366-1 Clause 3.27) is the process to explore or assess user interactions with the user interface. This may include, for example, usability tests, expert reviews, design audits, or cognitive walk-throughs. See Section 2.3.7 for further discussion of the evaluation plan.

1.5 User interaction

You have built the interface, you have ferried users over, and yet nothing happens? There is no spark or interaction of any sort. In other words, the movie is Gigli where Ben Affleck and Jennifer Lopez have yet to figure out their coexistence, and the audience suffers as a result. Just as in make-believe, understanding the interaction between your various actors – or perhaps users and the interface in this case – is the key ingredient to animating your interface. Modeling user interaction is the first step in gaining that understanding.

There is a plethora of tools and models available for modeling the use of your device and the interactions you see taking place using its user interface (UI). If the flow diagrams of the likes of the unified modeling language (UML) are not your cup of tea, you can use the popular approach of task analysis (see Sections 1.11 and 3.4) to shape your scenarios into stories and actionable observations. The two tools are not mutually exclusive, either. A discipline like service design may also be tremendously helpful here (see Juuso & Seppänen 2023).

We will discuss task analysis a bit later (see Section 3.4), but here it is enough to understand that IEC 62366-1 is all about ensuring safe use of the interface. Thus, interaction is modeled with the intent of identifying possible

Figure 1.6 A dearth of interaction can signal a lack of interest or a lack of clarity.

safety issues in the use of the device and the potential harms associated with these. For that reason, the standard, in Clause 3.11, introduces the term 'primary operating function' as a function that involves user interaction related to the safety of the medical device (see Section 2.3.2). This is noted to perhaps consist of a series of tasks, which themselves may comprise a series of user interactions. The somewhat vague term is given another definition in annex A3.11 of the standard where it is clarified as a function that is directly related to the safety of the device, as identified in some relevant medical-device safety standard. Such features are to be considered as input for the usability engineering process and observed appropriately in all subsequent work on hazard-related use scenarios. Prior to the 2020 amendment to IEC 62366-1, the annex also provided examples here including setting parameters for alarms, x-ray exposure levels, infusion levels, or even the series of displays the user is expected to navigate through. The body of examples appeared to range from explicitly critical properties to implicitly important attributes, which is perhaps why the list has been removed in the 2020 amendment. In any case, the gist here is that safety-critical functions deserve more attention than functions of lesser acclaim.

The term 'primary operating function', or POF for short, makes us think of devices with perhaps both a primary and a secondary mode of power. For example, a second world war submarine mainly intended to run on diesel engines, but occasionally switching to battery packs for silent running. This, however, is not what is meant by the term in standards. We feel that some other term, such a 'safety-critical function' might be more appropriate and more readily understood here, but the present term is what is used by all the related standards out there. Regardless of what we call such functions; they are functions we should take a hard look at with our safety goggles on. In cinematic terms, a primary operating function could be the module detach switch – or, more accurately, the detachment action itself initiated with a press of the button – in the Apollo 13 movie that gets covered up by a post-it note to prevent accidental use.

In practice a POF is a defined term and in wide use but not necessarily always understood in exactly the same way by all manufacturers. The differences we've encountered over the years have varied by both manufacturer and the type of device manufactured, but not to a serious degree. What matters here is that we as manufacturers have identified the critical functions implicated in the intended use of our device and, as a next step, have ensured that there are no usability issues decreasing the safety of the product in use.

For further discussion of primary operating functions see Section 2.3.2.

1.6 Use errors

The blame for the sinking of the Titanic is fairly universally placed on the shoulders of its captain, as depicted in Figure 1.7, *but stopping there would have prevented the discovery of other flaws in the design, construction, and operation of the ship. No similar broader lessons have been drawn from the modern-day grounding of Costa Concordia by its captain, so occasionally use errors really can be user errors, and as such out of the control of usability standards.*

Not all forms of incorrect use can be foreseen or controlled by the manufacturer – the IEC 62366-1 standard admits as much in its introduction – but nonetheless the errors can't be systematically brushed off as user errors made by the user. This much is also attested to by the standard in its annex A where a rationale for moving from talking about user errors or human errors to use errors is made by leaning on the design of the user interface as a source of many such issues. Such interfaces may be non-intuitive, counter-intuitive, or just plain hard to learn, all of which increase the probability of errors occurring in their use. The remedy, too,

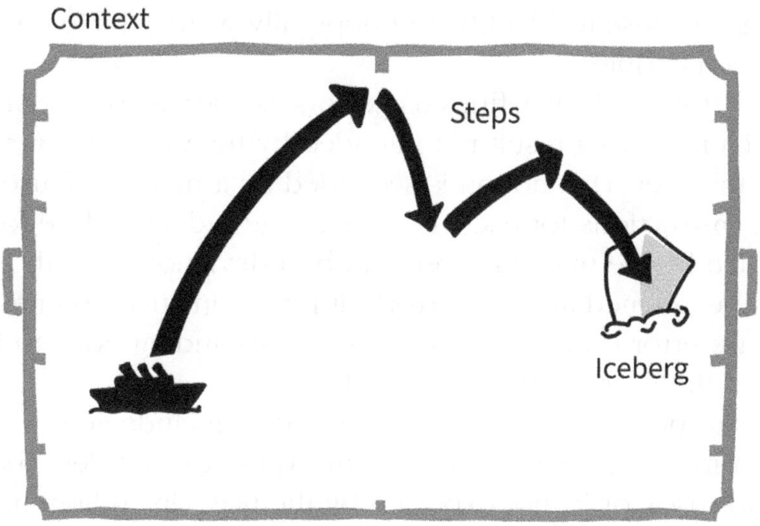

Figure 1.7 Unless Titanic had a magnetic pull towards icebergs, the captain's actions could probably be deemed user errors.

may be easiest to implement at the point of failure during the design process.

Such issues are further compounded when the user is acting under stressful or hectic circumstances such as are to be expected in emergency care contexts. When the planes hit Pearl Harbor during WW2 the first thought going through the first responders' minds probably wasn't to locate the instructions for use or the operating instructions of their medical devices. When reflexes and intuition govern actions all the information for safety in the world won't do much good, which is not to say that such information wouldn't have its uses in other less-heightened situations. Similar sad anecdotes can be drawn from prominent airline crashes of recent decades where pilots have perhaps been busy going through emergency checklists and not been able to notice or correct the true issues at hand. The correct remedy may have been on line 100 of the emergency checklist, but the plane had crashed by line 80. At a recent meeting of the International Medical Device Regulators Forum (IMDRF), it was reported as an anecdote that an intensive care nurse has to respond to an alarm given off by a medical device on average every 90 seconds. This is clearly not conducive to a calm, collected, and considered way of working where device instructions could be carefully consulted whenever needed. Murphy's law dictates that whatever can go wrong will eventually go wrong so it makes great sense to identify and hopefully ward off use errors to the greatest degree possible.

IEC 62366-1 Clause 3.21 defines use errors as user action (or lack thereof) which leads to a result not intended by the manufacturer or expected by the user. This notion is seconded by a note to Clause 3.3 where a) the instructions for use (IFU) are suggested as a place for defining results intended by the manufacturer, and b) a deviation from the IFU is suggested to be deemed as a use error if it results in such an unexpected outcome. A use error is also remarked, by a non-binding note, to include the inability of the user to complete a task.

The events to be interpreted as use errors thus include actions leading to a result a) not intended by the manufacturer (perhaps not described in the instructions for use), or b) not expected by the user (including the inability to complete some task). The reason for a mismatch between the result and the expected outcome may stem from, for example, characteristics of the user, the user interface, task, or the use environment. Note, however, that a malfunction of a medical device, or an unexpected physiological response from the patient are not considered use errors. Thus, a calculation error in the software, for example, is not a use error per se, but it may subsequently cause a use error as the user acts (or does not act) based on the erroneous information displayed by the device.

Similarly, pure errors in perception (e.g., user misreads a display) and cognition (e.g., user misinterprets the meaning of a number or a textual term) are considered contributing factors to a use error, not the use error itself. In other words, the standard links use error at the point of action (or inaction) during interaction between the user and the device. This is discussed in annex A3.21 of the standard. There are, of course, borderline cases where, for example, the bevel of a recessed segment display distorts a figure making it more likely to mistake a "5" for a "9". In such cases the solution should never be to instruct the user to be more careful when the design of the device can be improved to lead to a safer product.

Finally, the non-binding annex to the standard, A3.21, strikes an intriguing point of contrast between use error and risk. While with a risk there is always some potential for harm, a use error may be benign. A use error, such as Buster Keaton or Charlie Chaplin walking under the façade of a building falling over on a movie set, may not lead to much harm if a window or a doorway is fortunately positioned on that façade. Reducing those use errors that can result in hazardous situations and harm is the focus of the IEC 62366-1 standard. Note also that some flavors of risk management outside the domain of medical devices may also see risk on a spectrum from negative to benign to even positive, but in medical devices risks are – at least for now – always related to harms and not happy accidents.

1.7 The many causes of use errors

Bruce Wayne became Batman because his parents were killed by a madman, right? Or did he become Batman because he fell down a well inhabited by bats? Or because he had social issues and wanted to live by night? The relationship between cause and effect is not always immediately clear, but there can often be more than one path to some event as Figure 1.8 *illustrates. The same is true for use errors.*

One of the better contributions of the IEC 62366-2 guidance is a brief table in its clause 16.3 linking use errors and user-interface design issues together. This table contains a dozen simple examples showing what might be the cause behind a use error such as a user pressing the wrong button. Where the table falls short is that it only presents a single possible fault behind each error. In reality, the causes may be varied and will often also be layered, which is why a technique such as root cause analysis is often

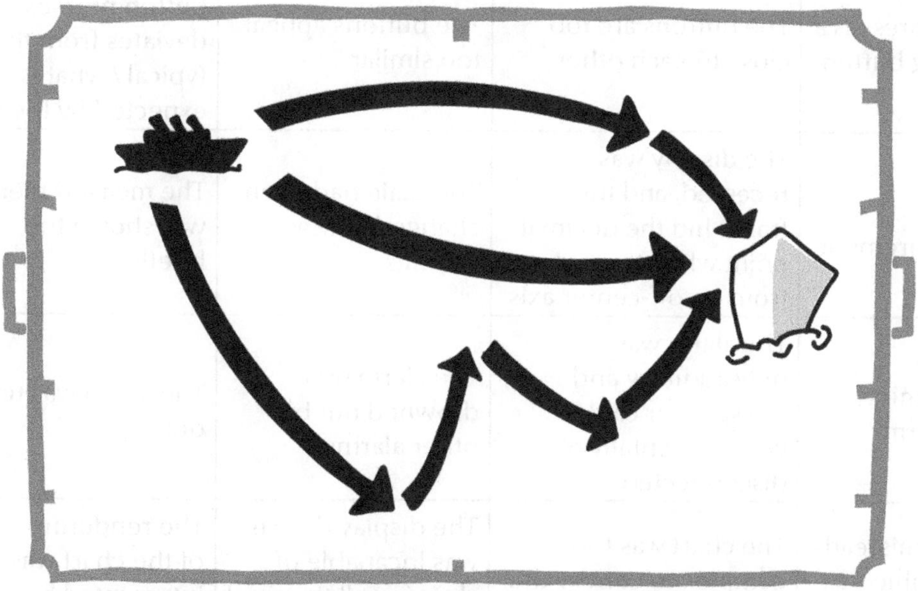

Figure 1.8 Not all roads lead to ruin, but more than one may.

used in quality management to investigate problems (see Juuso 2022). Table 1.1 shows a few possible UI-related explanations behind identified use errors.

The examples given here in the table are by no means exhaustive either. The point of the examples given is to prod us to look beyond the obvious or the first explanation we might think of. Oftentimes the simplest answer is the correct one, or at least the most commonly encountered one, but in our analysis of use errors we probably should not stop there. As with the Titanic example recounted in the previous lesson, stopping with the first plausible explanation (e.g., the captain was drunk with invincibility) would not have shown us all the flaws in the design and operation of the unsinkable ship that sank.

In the end the point of the previous story is not that the Titanic sunk, or that the captain should have been at the helm for two weeks straight, or that the waterproof bulkheads should have gone all the way to the top of the ship, or that the iceberg lookouts should have had binoculars – but instead we should have predicted at least some of these issues and

Table 1.1 Examples of use errors and possible design issues

Use error	*Possible UI design shortcomings*		
User presses a wrong button	The buttons are too close to each other	The buttons appear too similar	The sequence of button presses deviates from the typical / what is expected by the user
User misreads a measurement value	The display was recessed, and the frame hid the decimal point when viewed from on off-center axis	The scale had been changed in the settings	The measurement was shown too briefly
User misses an alarm	The alarm was only auditory and the speaker cable was accidentally disconnected	The alarm was drowned out by other alarms	The alarm was timed out
User misreads a graphic chart	The chart was too complex to convey the information clearly	The display device was incapable of showing all the used colors correctly	The rendering of the chart was interrupted by a program error

because wenow know about the issues, all future ships are designed appropriately. No risk assessment will ever be complete or valid for all time to come. The circumstances around a device, its use, and its users will evolve over time. Some risks will become less probable and others more probable. The expectations of the users will evolve even if the device itself does not, and thus not only should we not stop at the first potential cause identified but we shouldn't think that the causes identified will stay constant over extended periods of time. In medical devices, also the healthcare processes around the device may change and thus lead to changed usability expectations. Some of these changes may occasionally call for changes in the design of the device itself or its accompanying documentation.

1.8 Responsible organization

Upon first hearing the term 'responsible organization' it sounds like a euphemism for the Lee Harvey Oswalds of the world, or targets at the crosshairs of blame as Figure 1.9 *illustrates. It is not that. In fact, no specific responsibilities are placed on the responsible organization by the*

Figure 1.9 The crosshairs are on the entity accountable for ensuring the safe use of the device.

usability standard, but it is implied that this might be an active umbrella organization of users.

IEC 62366-1 Clause 3.12 defines a 'responsible organization' as the entity accountable for the education, training, use, and maintenance of a medical device (or combination of medical devices). This is noted to refer to, for example, a hospital, a clinician, or a lay person. Note that in the context of home use, a lay person may refer to the user or the patient themselves which may dramatically affect how the intended user is to be addressed during usability engineering.

Other clauses of the standard invoke the term only sparingly. In fact, it is only Clause 5.8 that mentions the term once in the binding text. The clause implies that it is the responsible organization who trains their own users based on the training and material they in turn have received from the manufacturer. Aside from this, the term is only used by non-binding notes and annexes. Via such soft language we learn that the manufacturer should ensure appropriate resources are available from them for the expected lifetime of the device. After all is said and done, the standard does not explicitly place any requirements whatsoever on the responsible organization.

The definition of the responsible organization appears quite toothless in the standard, but it does set the scene for expecting the users to also take some responsibility over their own use of a medical device. This includes education, training, and maintenance of the device. The toothless expression of this is understandable as most such organizations will likely not have read the standard, but manufacturers should have read it. All the responsibilities are therefore written for the manufacturers, but the expectations on the user organization are also softly voiced to help set the stage. The manufacturer won't be able to play hide and seek here as they are still on the hook for providing what is needed to the responsible organization, but the term may provide some practicality for organizing education, training, and maintenance for the devices.

At the very least putting the qualifier "responsible" in front of the user organization means that no cats should be put into microwaves and the fact that coffee in the takeaway mug is hot should not really come as a surprise to anyone. It also means that the user organization should not expect medical devices to be flawless or self-explanatory. This also means that the instructions for use that are included with a medical device should be read and adhered to, even if some organizations today place such manuals in the janitor's office in their original shrink-wrap. Another example will be that of a speaker cable coming lose during some cleaning performed around the device, and thus preventing audible warnings from sounding, which is not just an issue for the manufacturer.

Although the usability standard does not place requirements on the user organization, other documents may do so. For example, the instructions for use of any given device may set out expectations on the education, training, use, and maintenance of the device and call on the user to meet these expectations. Even if an individual user may be distracted by the hectic pace of their work, the critical setting of their patients, and other factors narrowing the breadth and the depth of their vision, the organization they work for shouldn't really claim similar extenuating circumstances. The organization should be able to ensure that all the users under their umbrella are trained and capable of correctly using the devices, as defined by the manufacturer. The organization should also be able to follow the manufacturer's instructions for the maintenance of the device. If this is not the case, the organization shouldn't really be using the device in the first place. The moral here is that an organization that tasks Mr. Bean with the restoration of Whistler's Mother doesn't get to plead ignorance when that facelift ultimately goes sour, as is shown in that film from 1997. Therefore, the inclusion of the term 'responsible organization' is more pertinent here in the standard than what its use in the clauses would suggest.

Having the concept of a responsible organization does us no good if that organization does not take full responsibility in all the ways a manufacturer expects. At the same time, defining as a manufacturer, all the ways the organization should take control will do no good either if the organization does not know the expectations or is not capable of meeting them. It is all fine and good giving a blanket warning that the contents of the takeaway coffee cup may be hot, but if they are usually not that hot and are only once so scorching hot that it melts the cupholder there is probably more blame to go around than just user error (which doesn't actually exist as far as the standard is concerned, see Section 1.6).

Not all errors can be prevented, but an error in communication or the alignment of expectations should be high on the list of errors we want to prevent. This also applies to handling events such as software updates, factory resets, and periodic cleaning of a device done for the user. Keeping up with the maintenance program of a device is naturally a desirable state of affairs and the responsible organization taking care of such work, instead of the busy medical professionals working under their roof, is no doubt a good thing but in all these cases changes occurring under the hood, or otherwise unnoticed by the actual user, will be a source of risk. A release note, a notification, or a worksheet to accompany a processed device may go a long way to mitigating such risks.

1.9 High appeal does not equate to good usability

The monolith in Kubrick's 2001 Space Odyssey is certainly sleek and cool, just like the button in Figure 1.10 here, but what does it do and how do you work it, is an undercurrent to the whole saga. The result may have been much the same if humanity had been handed an iPhone at that point in its evolution: we would not have known how to use the device in our hands, but it probably still looked cool.

Perhaps the easiest mistake in usability we have seen made by developers over the years is to think that just because they know how to use the device they built surely the users will too. Not far behind this is the mistake of thinking that usability is on a good level, as long as the graphical user interface (GUI) has been designed by a graphics designer and is cool to look at. Even some big-name manufacturers out there occasionally put out interfaces that look more at home in a modern art museum than in a

Figure 1.10 Hype may get in the way of true usability.

hospital ward. The reality is that no one will love a baby as much as its parents: no one will spend as much time getting to know all the facets of a device as much as its manufacturer hopefully does. Some users might end up doing that, but expecting all users to go through all evidence all the time and plan the use of the device as meticulously as NASA did when it wanted to take a spaceship to the moon is complete folly. Most of us just want a thing done when we are also busy on a dozen other things; so unless we use a device all the time we will probably lean as much on our own mental models of how a device should work as on any training by its manufacturer on how the device should be used. In this context, a cool-looking device that hides its features under a snazzy design aesthetic is less of a pleasure to use than a more ordinary device that makes its workflow apparent to the user. A good example of this is the subway system in Denmark's capital city Copenhagen: all the glass, stainless steel, and minimalist design evoke a sense of pleasing high design. At the same time, you have to walk right up to an escalator to see which escalator goes down to the platforms (there are no signs) and aside from a single display at one corner of the platform you have to board a train to see where it is going. All this works well for commuters who are already set in their daily patterns, but for the uninitiated the experience can be a little disorienting. Going back to the adult diapers we spoke of in Section 1.1, we can't imagine many users not knowing how they should put on patients, but if the same pair of adult diapers was designed to be shown off during the Paris fashion week the story could be quite different. Safety is not a priority in high fashion.

Innovation is great, it is the way forward from 2001 to 20,001 and beyond, but something that has to be explained is always going to be tougher to learn and most likely also to remember than something which comes as second nature. Interfaces today probably shouldn't still look like they were made for Windows NT 4.0 in the 1990s, but the way they differ in look and feel ought to be the result of appropriate consideration and not just marketing decisions.

As stated above in Section 1.1, the whole point of usability engineering in the development of medical devices is to create safe devices. The fanciest, most eye-popping user interface or the sexiest packaging for the device are not factors in this consideration – except as possible sources of risks or implementation sites for safeguards against the realization of risks. This much is stated by the IEC 62366-1 standard in its introduction.

A non-binding note to IEC 62366-1 Clause 3.11 suggests that safety as a concept encompasses the loss or degradation of performance resulting in an unacceptable risk to the patient. This is said to include use errors that prevent the user from effectively using the device to achieve its intended use. This definition is not completely at ease with the definition of safety in ISO 14971, but it is not at direct odds either. In practice, the gist here is that the benefit as well as the risk ought to be considered when assessing the acceptability of harms.

The scope of usability engineering around a device may be broadened from the essential topic of safety to accuracy, completeness, efficiency, and user satisfaction. Manufacturers are free to choose the breadth of their usability focus, as best meets their ambitions and commercial goals, but for a medical device safety must always be the most important, overriding part of that approach. See Section 2.6 for further discussion on qualities other than safety.

One thing the monolith from 2001 and the iPhone do have in common is the simplicity and cleanness of the design. Both objects look like they have been held firm against a belt sander until the result resembled a bar of soap – or at least that was the case until protruding camera humps became the norm on cell phones. Similarly, the Google search bar has been hailed for its powerful simplicity and as a result copied from here to eternity. Simplicity may thus have much to do with good usability; certainly in the sense of Occam's razor and finding the simplest answer to any need. The less moving parts there are in a solution the less things can go wrong, but depending on the parts, things can still go horribly wrong. Simplicity may, thus, not always be a good measure of usability. A startling example we've heard on a usability forum is of a patient ventilator with a simple on/off toggle button. This may indeed be a simple user interface, but toggling the setting one time too many by accident may also kill the patient. To us this example sounds too horrible to believe, and we might jot this down as an urban legend, but it does show that low complexity does not automatically equal good usability.

The take-home message here is that you shouldn't go for a quick fix to usability when safety is the real objective, and not just sales or snazziness.

1.10 Process to understanding usability

Usability is not a checkbox-activity, nor should it feel like voodoo. It is the result and status quo maintained by a systematic process of usability engineering. Instead of magic, the standard wants to see usability engineering as a process from inputs to outputs as illustrated in Figure 1.11.

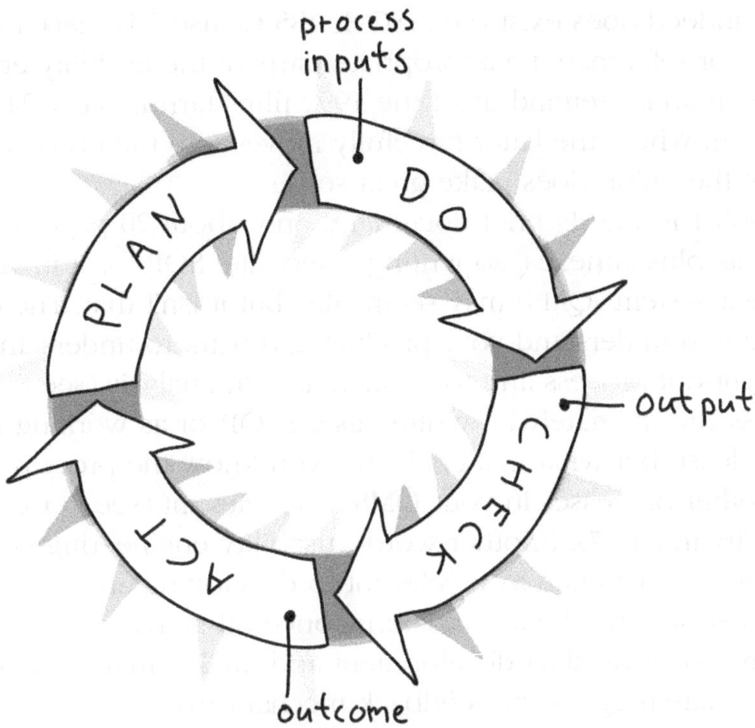

Figure 1.11 The Plan-Do-Check-Act cycle is a staple of understanding processes.

William Edwards Deming (1900–1993) is widely regarded as one of the founding fathers, if not the father, of modern quality management. He was an American engineer and statistician whose thoughts on organizing, measuring, taking responsibility, and improving production to attain higher quality of both products and processes have come to inform just about all activities in process management and the standardization of processes. IEC 62366-1 is no outlier here. Instead of a bag of magic tricks, or a book of spells, the standard wants to see manufacturers approach usability engineering as a systematic, documented process, tied to the appropriate places in their overall activities, and with relevant impact on the design, manufacturing, and risk management of products.

The key quality management standard, ISO 13485, does not force you to have a standard operating procedure (SOP) on usability, but if usability and user interfaces are topics relevant for your product you will want to consider developing a documented process here. However, the IEC 62366-1 standard on usability is not happy standing by as a mere wallflower here and interjects in its Clause 4.1.1 to demand that if a process for product realization indeed does exist (i.e., ISO 13485 Clause 7 is met) it must incorporate or reference the appropriate parts of the usability engineering process. Such antics remind us of the 1992 film starring Steve Martin and Goldie Hawn, where the latter forcefully moves into the house of the former, but the union does make great sense.

IEC 62366-1 is a fairly brief document, only about 20 pages of requirements plus annexes, so writing a separate SOP for it in your quality management system (QMS) may seem silly, but it isn't that. The expectation here is that you understand your product and want to understand its use via a thought-out process and tools such as task analysis (see Section 3.4). The process can ultimately be written as an SOP, or as working instructions at the very least, but what matters is that you know the process and that it links to other processes in your QMS as is relevant (see, for example, Sections 1.16 and 1.17). In other words, usability engineering is a process you will run and maintain as is relevant to developing and maintaining your products. It should not be a standalone, once-off activity but should instead be fused into your product development and quality management activities with appropriate triggers and feedback mechanisms.

1.11 Tasks, use scenarios, and the use specification

Some films, like Ocean's Eleven, may take a long time to set the chess pieces in place for a hidden solution to be operatically revealed at the end. Other films may hide the meaning of it all until the final curtain close or even leave this up to the viewer completely. Developing medical devices should, of course, not be shrouded in any such mystery. The purpose of usability engineering is to ensure that we all know where we are, what it is we want to happen, and that the steps to get there are not only predictable but also intuitive. These aims translate to the use specification, use scenarios, and tasks as illustrated by Figure 1.12.

IEC 62366-1 Clause 3.14 introduces the essential usability concept of a 'task'. This is said to consist of one or more user interactions with the medical device to obtain a desired result. A description of the task should not rely solely on references to functions or features provided by the device but should include allocation of activities and operational steps taking place between the user and the device. In other words, no user really sets out to marvel at a trademarked feature you make available, but they will have some goal in mind if they do interact with the feature.

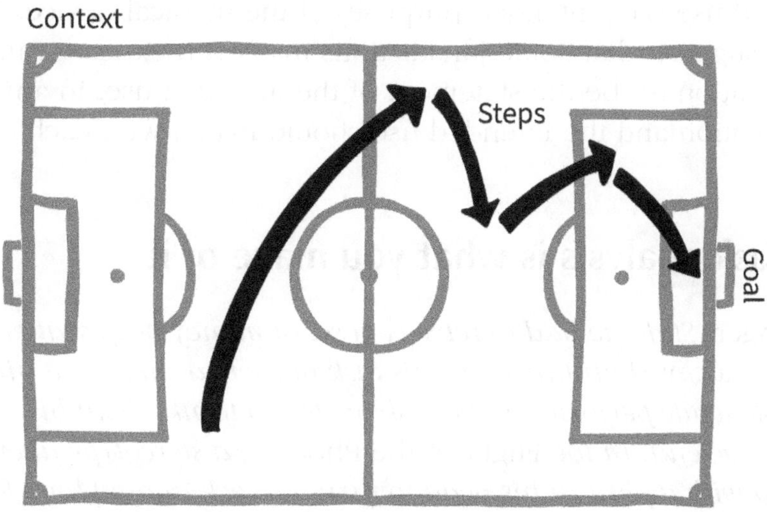

Figure 1.12 Use specification (soccer), a use scenario (score a goal), and the tasks therein (passes of the ball).

A 'use scenario' (Clause 3.22), then, is a sequence of such tasks performed by the user in a specific 'use environment' (see Section 1.13). As explained by the non-binding annex to the standard, A3.22, these scenarios can be defined in many different formats from story-like narratives to playlist-like definitions of tasks. The standard does not restrict you here, or provide you with a form to fill in. The core notion, however, is that the user wants to achieve a certain goal and undertakes a series of steps while interacting with a medical device to do so. You might define use scenarios for many different purposes during the design of your interface but remember that only those related to hazards and harms will get picked for in-depth scrutiny by the standard. The 2020-amendment to IEC 62366-1 further implies that tasks where a use error might lead to significant harm could be identified as critical tasks. This is not required by the standard, though.

Finally, the 'use specification' (Clause 3.23) is defined as a summary of the key characteristics related to the context of use for the medical device. This should include the use environment (see Section 1.13), but it is not limited to just that. In addition, the use specification (formerly known as 'application specification') may, for example, reference the medical indications, patient populations, intended body part or tissue type interacted with, intended user (including user profile), and operating principle. The specification made here should be regarded as an input for the intended use (i.e., intended purpose) of the medical device. A note to the clause suggests that some jurisdictions may even expect a summary of this specification to be the statement of the intended use. In any case, the use specification and the intended use should match with each other.

1.12 Task analysis is what you make of it

In Cliffhanger, *Stallone had to retrieve a set of money bags scattered around the Rocky Mountains, but his task masters didn't want him to be too comfortable while performing the task or stop to think about his chances of survival at the end. In the* Flight of the Phoenix, *a stereotypical engineer has to come up with a plan in his head to turn a wreck into a plane for there to be any chance of survival. Stopping to think about what it is that needs to be done and how is the essence of task analysis, as* Figure 1.13 *attempts to illustrate.*

'Task analysis' is a term tossed around by both the standard and the guidance, IEC 62366-1 and IEC 62366-2, as a potential solution to how you tackle usability. It may sound fanciful and impressive, but at the end

Figure 1.13 This task will self-deconstruct in 6 seconds.

of the day, it is exactly what it sounds like, no more and no less. It is not some magic potion concocted by the village druid that makes easy work of racking up the Romans until you have the N you wanted. It is not the monolith from 2001 that defies comprehension. It is not some sci-fi weapon or God-mode cheat code leading to tactical superiority in a first-person shooter like Doom, the game or the movie. It isn't the vastly complex supercomputer built in any number of films from The Hitchhikers Guide to the Imitation Game. It, quite simply, is the ordinary routine work you do in taking the extraordinary step into the shoes of your users.

It is ultimately up to you what lenses, tape measures, cattle prods, and far-flung assessment philosophies you use to understand your user and the tasks they face. The standard and the guidance attempt to give some sound advice on what to pay attention to during such task analysis, but neither will actually give you a ready-to-use set of forms or instructions. In fact, the standard implies that the guidance may do that, but even the guidance is mostly content in letting the vast body of undefined literature out there vaguely handle matters. We hope to stop the buck here, or more specifically in Section 3.4 where we dig deeper into the topic.

The way, shape, and form of task analysis is first and foremost up to you, but as a tool it should help you answer the sort of questions discussed in Section 3.4, it should work for your circumstances, and it should be fairly uniform from one project to the next. In this book we will present our model for addressing task analysis and recording the information in a practical cross-referenceable manner (see Section 4), but you should feel free to modify our approach as best meets your needs. There really is no single correct way of going about the work of understanding the user.

1.13 Formative vs. summative evaluation

The two terms in usability engineering you are most likely to hear after users and interfaces are formative evaluation and summative evaluation. These terms refer to explorative ventures into the safe use of your device (formative) and the conclusive, consolidative activities to verify that safety indeed is in reality, with all of the real moving parts, on a level you want to live with. The summative evaluation may sound like a sum of formative evaluations, as illustrated in Figure 1.14, *but different requirements apply to both types of usability evaluations.*

Figure 1.14 The Sum of All Fears may be a half-decent Tom Clancy novel, but it may not be the formula for successful usability evaluations.

The terms 'formative evaluation' and 'summative evaluation' may sound like just two more terms to get familiar with, but surprisingly enough, before the 2015 version of the IEC 62366-1 standard, the two actually used to be called 'usability verification' and 'usability validation', respectively. If you're familiar with verification and validation from other standards, such as IEC 62304 for software development (see Juuso & Pöyhönen 2024), this is probably a small lightbulb-moment that sheds welcome light on why you can't have your programmers pretend to be final users during the final summative evaluation. Verification is fine as an in-house activity, but for validation you always want real or at least realistic testing circumstances.

Present-day summative evaluations, thus, seem to be a direct descendant of usability validation, but formative evaluations today go a little beyond what you might expect of pure verification work: the goal here is to not only ensure we understand the use of our device, but also that we form that understanding through some usability activities. At the same time, formative evaluations may precede finished requirements (which would be necessary for verification work to take place) and may involve mockups instead of anything resembling a finished module or device. Validation thus is validation still, but verification is now appropriately called formative and sprinkled more liberally throughout our development activities from start to finish. Note that the objective in all medical-device usability activities is to ensure that the device can be used safely, not that the device is validated as everything the user has ever wanted.

We might loosely compare formative and summative activities to what the early explorers such as Christopher Columbus and Amerigo Vespucci may have gone through: go off on planned explorations of uncharted parts of the world, and once successful return to draw up maps of a passage to a new world or around the globe. The individual expeditions were formative activities and the consolidation a summative activity. Unlike in the case of usability engineering, the summative part in this example was largely derivable from the formative work when in usability engineering the two are often separate but related activities. The goal of the explorers was probably not defined as safety first either, or they would not have left their home ports at all. Like usability engineering, though, the goal was to learn something outside the foreseeable and beyond their own sphere of everyday experience.

In usability engineering, the formative evaluation might more realistically be done using a rubber dinghy to first paddle to a nearby island before attempting the fantastical voyage. Equally plausibly, it might just consist of showing post-it notes to the queen of Spain. The ultimate, summative evaluation might then be where the ship is manned by the whole crew, all provisions are bought, and the major stakeholders are lining the dockside as the ship lifts sails with great expectations. More ships would then follow if the voyage was a success.

Whimsical examples aside, the two key types of evaluations in usability engineering thus are:

- **Formative evaluation**
  This is defined in IEC 62366-1 Clause 3.7 as exploratory evaluation conducted with the goal of identifying strengths, weaknesses, and unanticipated use errors regarding the user interface. This is noted to usually take place iteratively throughout the design and development process (see Juuso 2022, Juuso & Pöyhönen 2024). Technically, formative evaluations may not always be necessary in the case of every usability engineering project or incremental change made to a product, but their use should always be addressed and a rationale provided. Also consider that formative evaluations may not be an absolute prerequisite for performing a summative evaluation, but at the same time forgoing formative work may increase the risk that a summative evaluation will not ultimately pass. The situation is similar as with clinical investigations where a pivotal investigation may be preceded by a pilot investigation. For more information on formative evaluation see Sections 2.4.1 and 3.12.

- **Summative evaluation**
  This is defined in IEC 62366-1 Clause 3.13 as evaluation conducted at the end of user interface development to gain objective evidence that the interface can be used safely. Note that despite its somewhat deceptive name a summative evaluation is not a summary of formative evaluations, instead it is the ultimate usability evaluation where after all the rehearsals everything of value is now at stake. The evaluation is noted to relate to the validation of the safe use of the user interface, but crucially it is not intended to replace the clinical validation of a medical device. In another note, this time to Clause 3.27 on user interface evaluation, it is stated that summative assessment is part of the verification and validation of the overall medical-device design. Note that the definition here does not call for the elimination of all risks, nor does it impose a definition of the degree of freedom from unacceptable risks (i.e., how safety is defined). The usability standard expects to see a single definitive summative evaluation for a medical device (see Sections 2.4.2 and 3.13).

Both types of assessment may consist of 'usability tests', which are defined as tests on the user interface in the context of particular intended users and an intended use environment (IEC 62366-1 Clause 3.19). The 'use environment' (IEC 62366-1 Clause 3.20), then, is described in terms of the actual conditions and settings in which interaction takes place. Examples here may include hygienic requirements, frequency of use, location, lighting, noise, temperature, mobility, and localization. The 2020 amendment to IEC 62366-1 adds social attributes (including individual v. teamwork, chaotic v. calm circumstances, stress and shift length) to these examples.

In terms of the use environment, it may also be necessary to look at further applicable standards from the IEC 60601-1 family of standards. For example, if your intended use environment extends to emergency medical care, IEC 60601-1-12 will step up to offer its thoughts on what the requirements for basic safety and essential performance should look like in this context. Examples here include additional requirements for switching a device on or off, the expected operating temperatures, the role of unexpected motion, the potential for confusion in a high-pressure context, exposure to biological materials, and various other risks observed in the use of the device. Similarly, IEC 60601-1-11 will want to be consulted if you are addressing the home care environment. In general, these standards point out relevant factors and useful considerations for both usability engineering activities and the eventual design and development input activities of a device.

1.14 Usability evaluation of accompanying documentation

Nokia used to be the greatest mobile phone maker in the world with billions of devices in use simultaneously. The Communicator phone they made was unique in many ways, but one of those was that you held it screen-outward to your ear. Except in movies. We have yet to see a film where the phone is used facing the correct way, thus indicating that the filmmakers did not read the instructions for use. As Figure 1.15 *illustrates, there are occasionally compelling reasons why users perhaps should.*

One of the top mistakes to make regarding usability engineering is that it is all about the graphical user interface of your device. Not so. The IEC 62366-1 standard wants to see you address all the buttons, pulleys, levers, hoses, and hatches your interface may have whether this is all software or has some physical parts to manipulate. The standard also expects to see your accompanying documentation be included in this although the recommendation to do so is notably only stated softly in a nonbinding note to Clause 3.26 (see Section 1.4).

The above naturally leads to the question: do all accompanying documents need to be available for the usability evaluations? No, but the documentation you claim is required to use the device safely, more than likely does. If you claim that 150 hours of flight training and reading an epic instruction manual is needed to use your device correctly that absolutely needs to be given appropriate standing during your usability engineering activities. You must also ensure that this training and material is both available and observed later in the real use of the device (although see Section 1.8). If you say that studying the onboard menu and choosing different meals with your copilot may be beneficial to ensuring a smooth flight but is not required, that is a different story. In other words, the choice of how you address accompanying documentation will depend on the role that the documentation has to play in the use of your device.

How about the various language localizations you may have of such documentation? Should those too be individually tested? The answer here again is that it depends. If the material is required for safe use of the device,

Figure 1.15 It's easy to dismiss the instructions until you realize you need them.

probably. If the material only acts as a tool of convenience, perhaps not. In less critical situations you may be satisfied with verifying the translation against the original version, perhaps by consulting a medical expert who speaks both languages. In more critical contexts you may decide to do additional testing using the translated versions in the appropriate settings. We might, however, argue that just rerunning usability testing is a poor substitute for ensuring the translations are of high quality in terms of both language and substance, and also appropriate for the indicated medical audiences. You might call this heresy, and it is that to some extent, but as usability tests are fairly complex situations you will probably not end up covering your documents fully or appreciating all possible misinterpretations of the content. Therefore, the usability evaluation of translated documents may be a good candidate for performing expert desktop reviews (see Section 3.11) instead of, or in addition to, summative evaluations.

The above should not be interpreted as a call to forgo usability testing of content intended to be translated. Usability evaluation should cover accompanying documentation in an appropriate way, and in as many languages as is deemed appropriate. Occasionally we have come across usability issues that may come down to not having enough usability testing that has taken place for a translation. We have, for example, seen a medical device routinely used at a pediatric intensive care unit which intermittently woke the kids up for no apparent reason by sounding an alarm with an incomprehensible error message. The alarm was translated into the local language, but it made no sense to any of the users and there was no way of reacting to the alarm or preventing it in the future. The alarm was no doubt programmed in good faith, but it ended up disrupting the sleep the patients needed to recover. If the alarm had been appropriately translated the issue could have perhaps been easily resolved, but as it was not, the issue persisted. We might argue that the issue here should have been caught during usability evaluations in any language: the issue was not just that the textual message was gibberish but also that a seemingly unimportant event – after all no adverse effects were observed along with the alarm – kept causing a real issue for the patients by denying them sleep.

The above issue could also perhaps have been resolved if the user was able to switch the error message into English, or some other language they understood. It may also be increasingly foolish to expect that the medical staff in any one country speak a language commonly spoken in that country. The world is no longer as monolingual, static, and neatly divided by country borders as it may have been in the 1960s, if it was that even then. The reality is that medical professionals come from all around the world and speak all sorts of languages as their native tongue. Providing a context-sensitive way to switch the language may be an appropriate design consideration in the future, and providing an easy reference code in addition to the translated message should really already be the norm. The simple context-sensitive IDs may not explain themselves to the uninitiated, but they will give us something to latch onto and investigate further should the main message escape us. After all, a lifeboat with no ropes around it to act as handles for pulling yourself in will leave you to drown.

Finally, symbols and icons are often brought up as a language-agnostic way to tackle labeling needs so that all users regardless of their language are equally at ease in understanding the labels. The issue we have with this seemingly smart solution is that the icons can then be equally indecipherable to all users regardless of their language skills. Symbols based on international standards (e.g., ISO 15223-1) are hopefully immune to such issues, but in the case of some other symbols we really should not take it for granted that the users interpret our scribblings in the same way as we do. In fact, we have occasionally wished the labels on some home appliance, such as a washing machine, were in any language – even those we did not speak – rather than the invisible language they happened to be in just because there was some strange drawing in the place of text. To say this is borderline heretic, we admit, but in our view common sense instead of magic should be applied every single time to every single design consideration. Magic may help, too, but it can't trump common sense.

1.15 Usability evaluation and clinical validation

Laurel and Hardy make for an entertaining duo, but would you ever mistake one for the other? Like the duo, usability evaluation and clinical validation frequently get rolled into one and treated almost interchangeably although they really should not be understood as synonyms. One is about safe usability and the other certainly more about efficacy and achieving an appropriate fit to its care context than just about intuitive interfaces. Figure 1.16 *illustrates.*

Usability evaluation often comes up for discussion at the same time as when a manufacturer starts to consider the state of the clinical validation they have for their device. The two terms are occasionally even viewed as synonymous, but this is akin to jumping the gun and attempting to spurt half-way down the track before anyone else takes notice. The two terms are related, but they do not mean the same thing.

Usability evaluation – in its summative evaluation form – takes the intended use of the device as given and proceeds to look at the hazard-related use scenarios around that use and possibly making the device more intuitively usable. Clinical validation, on the other hand, is primarily concerned with what the device is aiming to do and how convincing the evidence is for it to be able to accomplish what it was designed to do. Both

Figure 1.16 Both usability evaluation and clinical evaluation may be needed to pass a device to the market.

involve real users, representative versions of the products, and a series of evaluations, studies, or trials devised around investigating their respective goals. Both can represent roadblocks to allowing a device to move forward, but as checks the two are not interchangeable and do not focus on the same aspects. Years ago, we were being shown around the Sheldonian Theatre at the University of Oxford in England. On the private tour, we were amazed to learn that as part of the graduation ceremony all the professors from all fields of study at the university form a gauntlet that the new graduates have to walk through. If a single professor, from any field, stops a student from passing, they won't graduate. That is also how you should view usability evaluation and clinical evidence: as fellow gatekeepers of successive gates, or as Laurel and Hardy standing behind one another.

In theory, medical device development is mostly about ensuring safety, so would it not be enough to look at the safe use of the device? If all the risks were related to the use of the device, and no risks would be related to the functioning of the device; perhaps, in theory. Even then we would want to make a case for the benefit to be gained by using the device, and not just that it could technically be used or operated like another device of the same type. The bottom line here is that clinical efficacy can't be proven via safe usability, but a clinically effective device should also at the same time be safe to use. Risk management, too, will weigh in on any final green light based on both the benefits and the risks involved.

Historically, too, the oversight of medicinal products and medical devices has evolved from the wild west of the early 1900s, to ensuring safety, then ensuring efficacy, and finally as we approach the present day attempting to ensure the affordability of products. Not all products sold today meet all these qualities, but the qualifier pair of safety and effectiveness is now a commonplace mantra in the field. This also explains, in part, why usability, risk management, and clinical validation are such important activities in developing medical devices.

All medical devices are subject to proving their safety and efficacy, and thus expected to address clinical validation in some way. However, not all medical devices are expected to address usability evaluation as rigorously or at all. The underlying requirement for usability engineering lies in risk management and ensuring that devices are safe and can be used in a safe way by their users. If risk management indicates that use or usability – the action or lack of action by the user – may contribute to the occurrence of harm, usability engineering is probably indicated. If the risks related to such harms are unacceptable as per your risk management activities, the application of usability engineering is strongly indicated. In other words, if usability can play a part in the risks you have identified, you would do well to consider how usability engineering can help you achieve a greater level of safety for the use of your devices.

1.16 Risk management always runs parallel to usability

Clint Eastwood, a giant among moviemakers, is also distinguished by having made two films about the same events from two opposite points of view. Flags of Our Fathers *and the* Sands of Iwo Jima *exist on two parallel coordinate systems, but both come together around the same battles of the Second World War. In much the same way, usability engineering and risk management are intertwined, as* Figure 1.17 *illustrates, but admittedly fight on the same team.*

IEC 62366-1 expects usability engineering to tango with risk management. It does not specify which activity leads, but it does point out the steps to the dance. In Section 2 we will go through the required activities of usability engineering in detail and at several points throughout the processes you will note a link to risk management. From the point of view of IEC 62366-1 risks have a usability factor, an aspect where usability provides the decisive wind gust at the watershed point between a harmful and a benign outcome to a risk. ISO 14971 has field command over the wider battlefield of risks, but it too should be happy to receive assistance on the usability front. Perhaps Al Pacino is the wise, old ISO 14971 risk management and Gabrielle Anwar the beautiful woman in black who lends her eyes to Pacino in terms of IEC 62366-1 usability, just like in that tango from *Scent of a Woman*.

This marriage between risk management and usability engineering is not an oddity, or a once-of-a-kind occurrence in medical devices. In fact, risk management is connected to just about all processes in the design, development, manufacturing, release, support, follow-up, and decommission of medical devices. For those areas where risks are connected to the device and the users fairly directly, we rely on ISO 14971, and for other areas where risks are remote at worst we might employ a risk-based approach rooted more in common sense than ISO 14971. In many ways, risk management is an alternate view into everything we do, and everything we don't want to see taking place. It is a perpendicular line of sight into ISO 13485 quality management, a vision of the Voldemort we might be afraid of and at the same time the punctuation of weight that adds

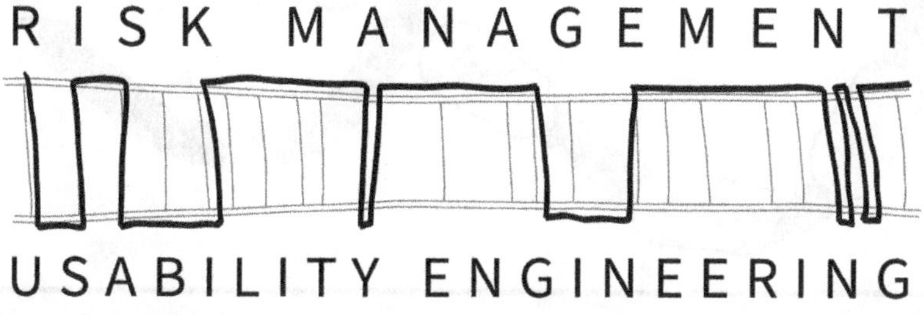

Figure 1.17 Risk and usability have a lot in common.

the relative levels of importance to everything we do. From the point of view of IEC 62366-1 usability engineering, it is of paramount importance that usability and risks each play their parts in ensuring the safety of the medical devices we make. Progress or activity on one front of the battlefield also leads to actions, or is at least noticed, on the other.

The link between usability engineering and risk management is discussed in more detail in Section 9.2.

1.17 Usability is the pace car for software development

Tom Cruise packed some explosive energy in the Days of Thunder, *and ultimately became a better driver once he learned how to work with the characteristics of his car, the track, and the expectations of his audience and teammates. This path to enlightenment took some dark turns and hitting his head on walls, but he got there in the end. Usability engineering can act as a pace car of sorts in our race to develop the best new medical devices out there, as* Figure 1.18 *illustrates.*

No software developers want to produce bad code; no piece of software is intended to be unusable. Good usability in software casts its sights wider than just the safe use of the software, but safety is what the IEC 62366-1 standard focuses on. IEC 62366-1 may be thought of as the specialist consultant brought in to ensure your designs or implementations are up to scratch when used by someone else besides you and your proud parents.

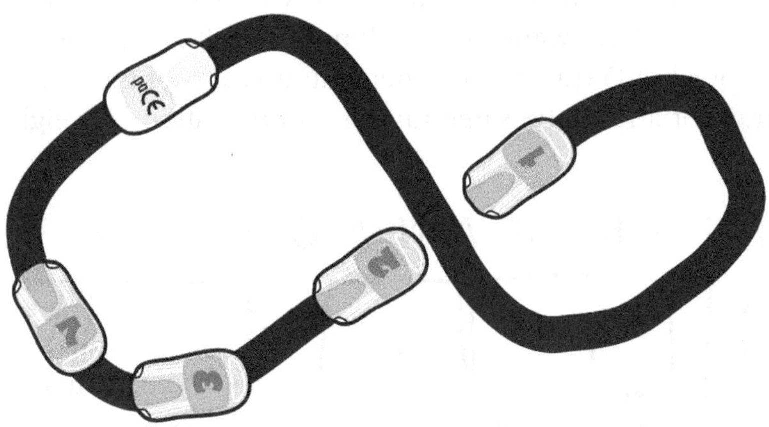

Figure 1.18 Just like a pace car, usability engineering may have a role at different points throughout your race.

Usability engineering ought to be a part of all your considerations all the time, but more often it is some late check or gate your software must pass through to get its passport for the wider world out there. It, however, does not have to be that.

The activities proposed by IEC 62366-1, with the possible exception of the summative evaluations discussed in Section 2.4.2, are just the sort of activities you should run before writing a single line of code to understand your intended use, users, and use environments. In this sense, usability engineering could feed into your software development activities via, for example, testing out ideas, needs, paper prototypes, and mockups of various aspirations. Many of the key documents expected by usability engineering (see Section 2.3.1, for example) could be extremely useful when you start defining the requirements of your software and the subsequent design documents. Even then, you would still want to hold off on performing the summative evaluation until the implementation is mature. You might, however, want to keep the dialogue between usability engineering and software engineering going throughout the development project, as starting out early with usability might give you a better product and less costly issues to fix along the way.

In our two earlier books we discussed ISO 13485 quality management and IEC 62304 software development, and how the two might be negotiated together in a practical, lean way. The same should be attempted in terms of IEC 62366-1 usability engineering as well. The basic frame of thought here we would suggest is that usability engineering is the pace car that can begin, punctuate, and even finish our race to develop safe medical devices. The point is that usability engineering fits into several roles and at several points throughout our product development work, and it will depend on the circumstances of your projects what makes the most sense. We will discuss usability and product realization in greater detail in Section 9.1, but also in every section of this book before then. For a discussion of how usability engineering can inform our work from a concept to a finished product see Section 10.

1.18 User interface of unknown provenance (UOUP)

It's not quite an Oops, but UOUPs are a pullback to the past just the same. UOUPs, as illustrated by Figure 1.19, *are the nomads who wander into our present-day context without much history. They may be necessary to save the*

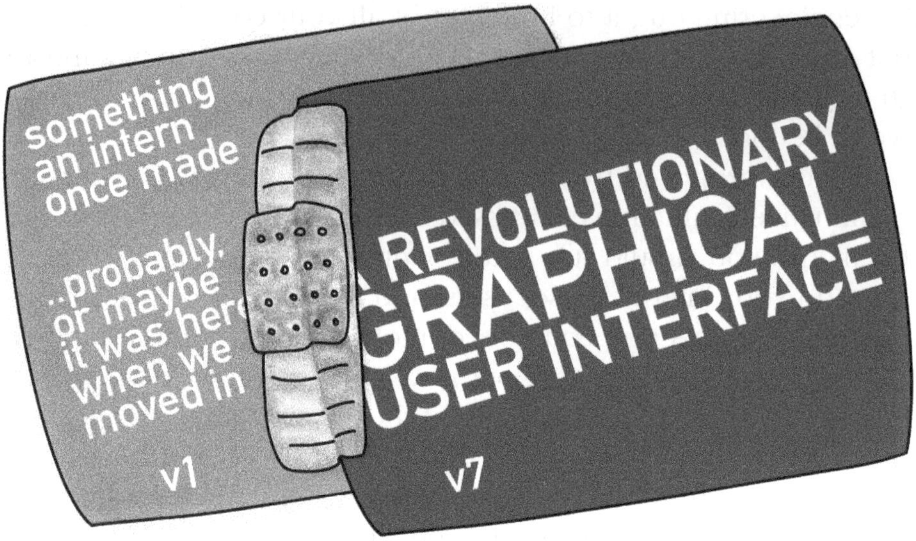

Figure 1.19 Sometimes the ghost in the shell can also lurk between the shells.

day, just like that man without a name immortalized by Clint Eastwood, but they don't wear their birth certificates on their sleeves by any means.

Medical-software standards such as the IEC 62304 (see Juuso & Pöyhönen 2024) define a term for software of unknown provenance (SOUP) to address software created through a process we don't have visibility into, full information of, or control over. While IEC 62304 also defines the term 'legacy software' to address previously developed software not yet meeting with the full expectations of the standard, IEC 62366-1 merges these two notions into a single concept of UOUP.

IEC 62366-1 Clause 3.15 defines UOUP as a medical-device user interface (or its part) which has been previously developed and may be lacking in its records for the usability engineering process as expected by the full standard. In IEC 62304 parlance this might more accurately be called a legacy user interface, instead of something drawing upon SOUP (see Juuso & Pöyhönen 2024), but the intention behind UOUP is nonetheless clear.

Philosophically we might ask if there is such a thing as a user interface of unknown provenance? The code behind the interface may certainly be of unknown provenance (i.e. "SOUP" in origin), but the interface itself is something then released by the manufacturer of the medical device. In this sense the interface is never something just passed through the manufacturer but is instead the entirety of the user-facing side of the medical device, part of which may be put together from legacy parts or parts developed elsewhere. Note that also the information for use (including user manual) is a part of this same interface. With this in mind, the concepts of SOUP and legacy may somewhat comfortably be rolled into one concept of UOUP, as they have been by the standard.

See Section 2.5 for a more in-depth discussion of UOUPs.

1.19 Usability engineering file

Remember that warehouse full of wooden crates at the end of the Raiders of the Lost Ark*? The usability engineering file is that warehouse, but hopefully in a cross-referenced and easily searchable manner. What is meant here is actually a user interface into your medical device file, a way of pulling all essential usability-related records. The usability engineering file could be in*

Figure 1.20 The Ark of the Covenant should not go into your UEF, but a lot of other items will.

a silo of its own, but most likely you will want to reuse existing documents where possible and ensure you don't reinvent the wheel unnecessarily.

As you may have guessed from the above, the usability engineering file is an umbrella term for all usability engineering documentation. This may be one neat dossier, or it may be a network of all the necessary information contained in your medical device documentation. In our previous two books (Juuso 2022, Juuso & Pöyhönen 2024) we presented a network of documents expected from ISO 13485 quality management and IEC 62304 software development. All those documents may be linked to your usability engineering activities and documents. You might be replicating work if you now developed each one from scratch under the heading of usability. You might also be setting this all up in a silo and torpedoing any chance of meaningful information exchange between the processes, thus perhaps quarantining valuable information when it should be shared and reacted on.

Perhaps the prime candidate for repeated work, or the most prominent topic of mutual interest between usability and your other quality management activities, is risk management. Don't reuse your risk management to wave off doing activities in usability engineering but instead ensure that the two processes talk to each another as is needed: when new risks are encountered on either track you let the parallel track know, and whatever you do, you don't silo the results separately.

We would urge you to think of the usability engineering file more as a tag or a network of cross-referenced documents in your overall medical-device file than as a single document or a silo of documents. See Section 3 for a further discussion.

From a conformity assessment point of view, how you have handled your usability engineering activities will be assessed more through looking at the usability engineering file, than by just marveling at the product you roll out on the showroom floor. This is hardly surprising, but it means that if you happened to just stop the merry-go-round at the perfect time and ended up with scissors that don't cut off the user's fingers you will get called out on it if your process and documentation only amount to controls for an amusement park ride. This is true for just about all conformity assessment activities always.

1.20 How much usability engineering is enough?

K.I.T.T. may know the answer to this question, he definitely had the front light strip to match our Figure 1.21 *here, but maybe Michael Knight didn't*

Figure 1.21 Usability is not a one-size-fits-all, cookie-cutter activity.

ask him, and now we can't. The best answer is therefore that it depends, and that you are in the driver's seat until your choices are seriously called into question.

Doing nothing about usability engineering is plainly wrong if your medical device has some kind of user interface. Remember that a user interface does not just correspond to a graphical user interface but instead consists of all the interaction your users can have with your device and its accompanying documentation. This, as discussed in Section 1.3, can be a wide-ranging concept. Usability engineering may even be appropriate if your device does not appear to have any human users at all, such as might be the case with some machine-to-machine medical-device software library. In such cases you might need to consider the people integrating your device into their device as one type of user for your API-based interfaces.

Doing something, anything, about usability engineering may technically be enough for regulatory purposes. On industry forums you'll hear discussion of how notified bodies might call for usability engineering activities but once seeing some activities having taken place in a recorded fashion, be happy to move on. You may also hear stories of other notified bodies, or the FDA in particular, diving deep into your usability engineering file. They might even ask you poignant questions on how you ensured your definition of users and user groups was appropriate, and how these matched with the representative users you had take part in your summative usability evaluations. This will particularly be the case with high-risk devices and if your usability efforts drive your clinical evidence in any major way.

The complexity of your device, its intended use, and the risk associated with it will all influence the depth you will want to wade here. The more usability factors can have an effect on the risks or the benefits associated with your device, the greater the need to address usability engineering. On the other hand, a simple device may have grave usability issues that should be weeded out carefully. The sheer number of features may thus not always be a good measure to use here even if it can act as a sound indicator of the scale of the effort to be expected. A startling example we've heard on a usability forum is of a ventilator with a simple on/off toggle button. This may indeed be a simplistic user interface, but toggling the setting one time too many by accident may kill the patient. To us this sounds too horrible to believe, and we might jot this down as an urban legend, but it does illustrate nicely that low complexity does not automatically equal good usability.

Generally, formative evaluations are seen as an optional tool available to the manufacturer to use as they see fit, while at least one summative evaluation is regarded as necessary for addressing all the use-related risks identified for a device. In other words, a manufacturer might use formative evaluations to get to the point where they feel confident in running the final summative evaluation. Interestingly, the nonbinding guidance for the usability standard, IEC 62366-2, also leaves a door open to not conducting any summative evaluations in its Clause 6.7d. The implication here is that the usability engineering activities may be tailored to match the needs of the device and its usability, but that the argument should be appropriate to the risk and complexity of the device, and the significance of the change we are perhaps making in that device. Take this to refer to either new devices or device changes where in either case no usability-related risks are implicated.

Doing enough about usability may be a conversation with your regulatory authorities, but it should absolutely be a conversation with yourself. You address usability until you are certain enough that no unacceptable use-related risks remain, and until you feel you have extracted the other non-safety-related benefits from the activities you wanted. What is enough is, in other words, again up to you – up to you to define and up to you to argue in the context of the device you manufacture.

1.21 What should we cover in usability evaluations?

What to include and what to leave out is a question any film adaptation of a novel must wrestle with. It may be the deciding factor in either courting or staving off favor from die-hard fans of the original text in the case of films such as Dune *or* Lord of the Rings. *In usability engineering the similar question of what to include in the summative evaluations is less about the completeness of character arcs and much more about the safety of users.*

The amount of toiling over the script of the processions beforehand is likely similar though. Figure 1.22 *illustrates.*

Once you have decided that some usability engineering activities are in order for your medical device, the next question may be just what to cover in those assessments. This too is a sliding scale from the safety of the medical-device user interface to the perceived user friendliness of everything involving that interface, but the standard does expect to see a hard core in that body of work.

At a minimum, your usability evaluations might only cover the following:

- **Hazard-related use scenarios**
  The subset of hazard-related use scenarios you identified as critical for the safe use of the device. This selection and the criteria for it is to be documented, for example, in your user interface evaluation plan (see

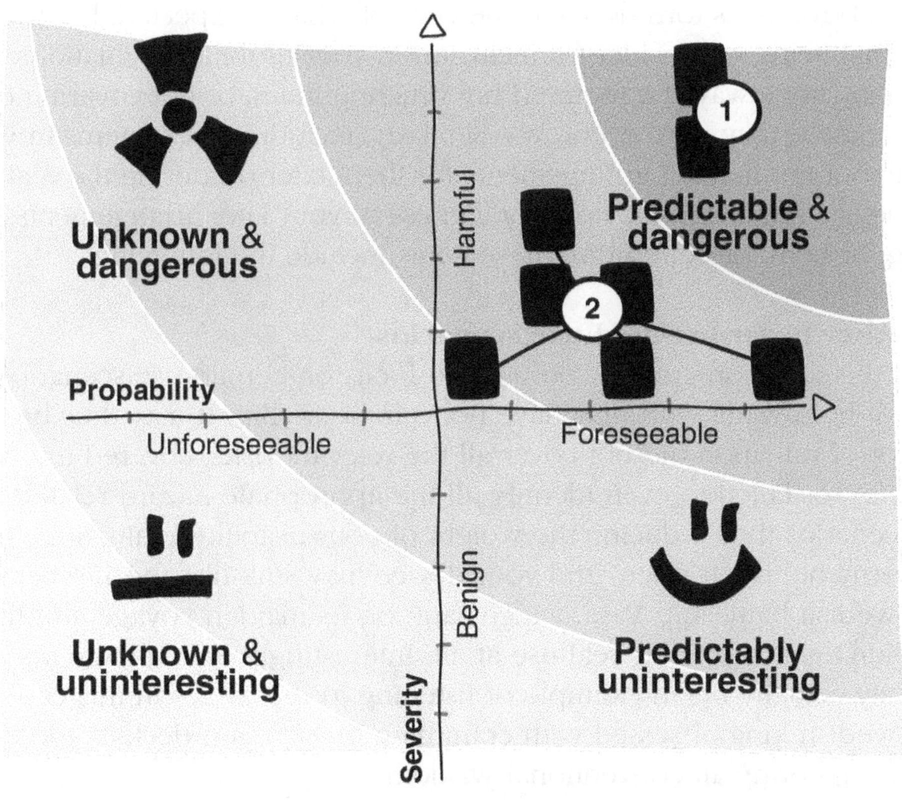

Figure 1.22 The required usability engineering work (top right in the figure) is perhaps only a quarter of the work we could be doing.

Section 2.3.5, 2.3.7). In the absence of good criteria, the expectation is that you cover all identified hazard-related use scenarios, and that you have identified all relevant scenarios.

- **Information for safety**
  The subset of 'information for safety' you decide to use in reducing usability-related risks to an acceptable level in your risk management activities.

Note that for later revisions of your device you might only address changes in your user interface, and thus cover the critical hazard-related use scenarios and information for safety around those changes. Usability evaluation is not, despite what might occasionally get said in casual conversation, a take-it-or-leave-it kind of an activity. What matters is that you address the safety of your device adequately and that your process allows enough practicality in this to not just lead to a perfunctory checkbox activity – or worse yet, one always postponed.

The above list is torn down to the core of what is expected, but there are additional items to consider for inclusion in your usability evaluations. These items may not always be required but omitting them, or not covering them appropriately, may raise eyebrows. Similarly, including these items in your testing, but not making arrangements for them later on during the real use of your device might call the appropriateness of your later arrangements, your testing, or both into question. These items include the following:

- **Other hazard-related use scenarios**
  If the selection criteria you used to focus on certain use scenarios does not hold water, you may have holes in your ship. It may thus be that your evaluation did not cover all the relevant risks, covered the wrong risks, or failed to even identify all the appropriate hazard-related use scenarios thus reducing the weight of your usability evaluation. Too many holes left open, and your device may sink like the 17th century Swedish battleship Vasa, which sank on its maiden voyage and thus didn't get to see any real use at all. Interestingly enough, the Vasa may actually be an example of listening to the user – in this case the Swedish king obsessed with cramming on two-gun decks – too closely and ignoring all conventional wisdom.
- **Instructions for use**
  Information for safety is specifically called out by the standard for inclusion in your evaluations, but also the wider body of instructional

material you provide may be important to consider when planning your usability evaluations. Not all types of devices are required to have instructions for use, but if such instructions do exist the expectation is that these make up a part of the user interface of your device (see Section 1.4). We would avoid a knee-jerk reaction to this realization and not require that all possible user-facing documentation is included in usability evaluations, but perhaps the documentation that features in your critical hazard-related use scenarios and information for safety should. Naturally the opposite is also true: don't include material you have no plan of including for later real use.

- **Training**
  All that we said above for the instructions for use also applies to the training you envision on the use of your device. We also feel that it would be somewhat dishonest to dope out your test audience with extensive training you know won't be there for real use of the device or have them cram just before performing complex procedures when you know that won't be the way your device will eventually get used in real life. Thus, overdoing training for the evaluations may diminish their value in assessing the real use of the device later. The standard does not get vocal on the matter so plan as best meets the needs of your device.

From the above discussion it should be clear that the purpose of usability evaluations is to bring up any safety issues with your interface and iron these out before releasing the product. In order to be successful in this, the evaluations should be as true to life as possible, which is why the design of the evaluations matters a great deal. In this book we will look at planning your usability evaluations in detail starting from Section 2.3.7.

In Figure 1.22 we have a quadrant labelled as "predictably uninteresting" which covers risks with more or less benign outcomes. It is fair to discard this from traditional risk management, but if we get serious about understanding the benefits around our device this might become an interesting space to look at. The ISO 14971 standard on risk management, of course, appreciates benefits, even if it mainly looks for benefits to go along with risks in the top-right quadrant of the figure. The benefits of a medical device are thought to mainly come from the clinical evidence built for the device, as in a benefit being the availability of some feature referenced in the intended use of the device, but the benefit could also come from how well the device fits into clinical practice or how much it improves that everyday practice. In the latter case benefits could, and should, be of great interest to our usability engineering activities – and not just in terms of preventing harms but also ensuring benefits. For more discussion of benefits see Section 6.

1.22 When should we revisit usability evaluations?

Why did it take Tom Cruise almost 40 years to resurrect Top Gun *for its first sequel? Why does Francis Ford Coppola constantly fiddle with his films like* Apocalypse Now! *so that each new re-release on some new format is always slightly different from the last? We don't know, but the answer probably somehow depends on the circumstances of each project.* Figure 1.23 *illustrates.*

The IEC 62366-1 standard doesn't go into great detail on when to redo usability evaluations. It does give us expectations on what to achieve by usability evaluations, especially through its discussion of the summative evaluations (see Section 2.4), and it does occasionally reveal intentions of enabling simpler methods (e.g., desktop evaluations, see Section 3.11)

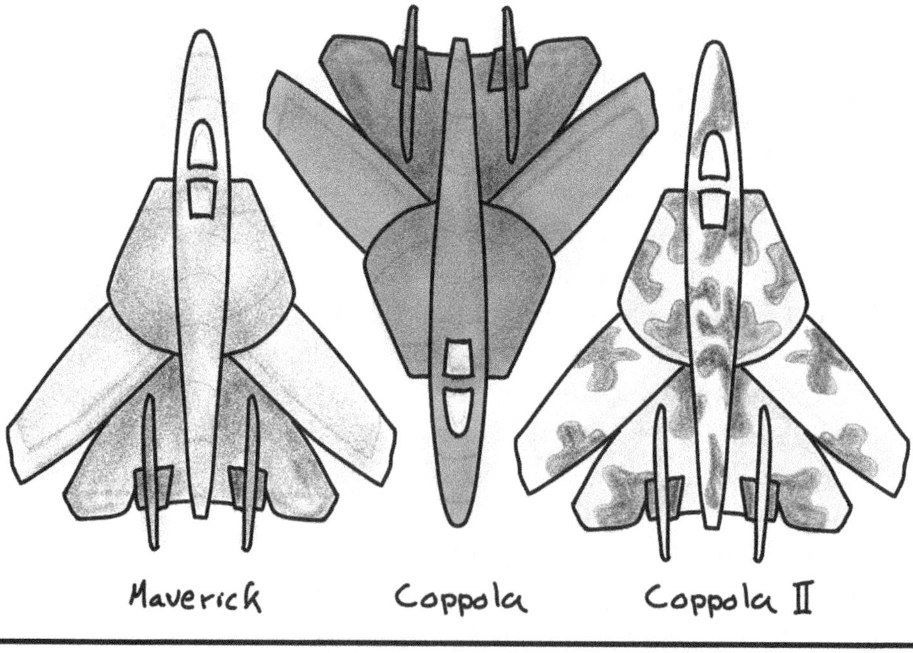

Figure 1.23 Only the shapeshifting Decepticons from the Transformers franchise may match Coppola's appetite for change.

for assessing the usability of later changes to the interface. It does not, however, give us any explicit discussion of when a lighter approach might be appropriate versus rerunning the full gauntlet. So, how do we know if a light desktop review is appropriate or if we should make arrangements for a whole new usability engineering evaluation?

In practice, the answer here too is that it depends. As an answer this is both infuriating and liberating as it is now up to you to analyze and argue what is appropriate. In our previous book, Juuso & Pöyhönen (Routledge 2024), we suggested scaling your response to proposed software changes by first assessing the scope of the change, then identifying its effects in your product and processes, and then proceeding as indicated by that scoping activity. A small, inconsequential change with no effects on your risk management or regulatory requirements should be much easier to execute than a change that can significantly affect the user and thus has implications for risk management and regulatory conformity. Here we suggest the same for addressing the question of revising the usability engineering file of your interface and device.

The only totally incorrect answer here is that you don't revise the usability engineering file because it was already done back in the day and now seems too laborious and costly to review or redo. Here are a few basic triggers we see as giving rise to taking a hard look at your existing usability engineering file:

- A new version of the IEC 62366-1 usability engineering standard comes out that adds discussion on the types of products or interfaces you develop.
- A new device-specific standard is published or added to your list of requirements that sets new usability expectations (e.g., collateral and particular standards in the IEC 60601family of standards).
- You are planning a full rewrite of the software user interface code, perhaps in a new programming language or for a new kind of device. Note that while the user interface specification might remain the same, the new programming language or the new software libraries used could bring subtle changes with potentially significant impact on the user (e.g., new size, spacing or behavior for form elements). In such cases, at least a desktop review may be warranted.
- You notice a trend in customer feedback indicating an issue in your user interface, software or otherwise. Remember also that the

accompanying documentation (e.g., instructions for use) are typically considered a part of your interface (see Section 4.4).

- You are overhauling your risk management file and may have risks related to the usability of your device.
- You suspect the use of your device has changed, e.g., due to new care practices or new types of competing products becoming available on the market.
- You want to add a novel new feature to your device and don't know how your users will take it. Remember that introducing a significant new feature under the hood, an autocorrect function for the angle of attack of a passenger plane, without telling the pilots, was also at the root of some much-publicized airline crashes in recent years. People died then and they might die again if your medical device cut too many corners.

The above list is nowhere near comprehensive, but perhaps it is illustrative of the telltale signs that something done once does not stay valid forever. The standard and the guidance, IEC 62366-1 and IEC 62366-2, do a good job of showing us the rules of usability engineering, but neither document should be considered as a travel guide for our trek from here to eternity. In fact, both documents appear fairly content instructing work leading up to our big product launch – whether this means addressing usability through full-blown usability engineering or some UOUP-infused process – but neither appears too keen to jump onboard and help us navigate the incremental course corrections we are virtually guaranteed to need over time. The notion of change control, or the role usability engineering should play in assessing changes to products, is not explicitly or comprehensively discussed by the standard or its related guidance. As a result, we fear that many manufacturers will balk at the price of redoing everything every time a change is discussed, and as a result opt to do nothing at all even when the right thing to do could be a small activity addressing the change itself.

1.23 Rapid versioning for development or deployment

The film Snakes on a Plane *is perhaps the foremost example of a movie its audiences got to shape before it was released – at least by introducing one obvious expletive-filled one-liner for its star Samuel L. Jackson. Like films, medical devices often take millions to develop, have great expectations to meet, and – unlike films – may even cause serious harm to their audiences if things go wrong. Thus, the plan to wait until the plane is in the air to find out just how many snakes snuck on board is not all that tempting, but yet most everyone is in a rush to dispatch those planes.* Figure 1.24 *illustrates.*

In the previous lesson we looked at when to revisit usability evaluations over time, but you might instead be looking to perform rapid prototyping during development or perhaps even issue new versions of the finished product in some rapid succession. How should we address usability engineering in these contexts?

For work taking place before product release, formative evaluations are a natural fit. Here you are free to execute your user interface evaluation activities as per your plan and can investigate issues and design ideas in the safety of your playground. The same also applies to summative evaluations, but there you should be more careful in your evaluation design to ensure you can build on the outcome as you hoped. In either context you can initiate a study with a new prototype as soon as you want, as frequently as you want, and on the scale you want – but in the case of summative evaluations (see Section 2.3.7) expectations apply if you want to then rely on that evaluation for later steps. Remember too that regulations will also have their say here if you come anywhere near the clinical use environment.

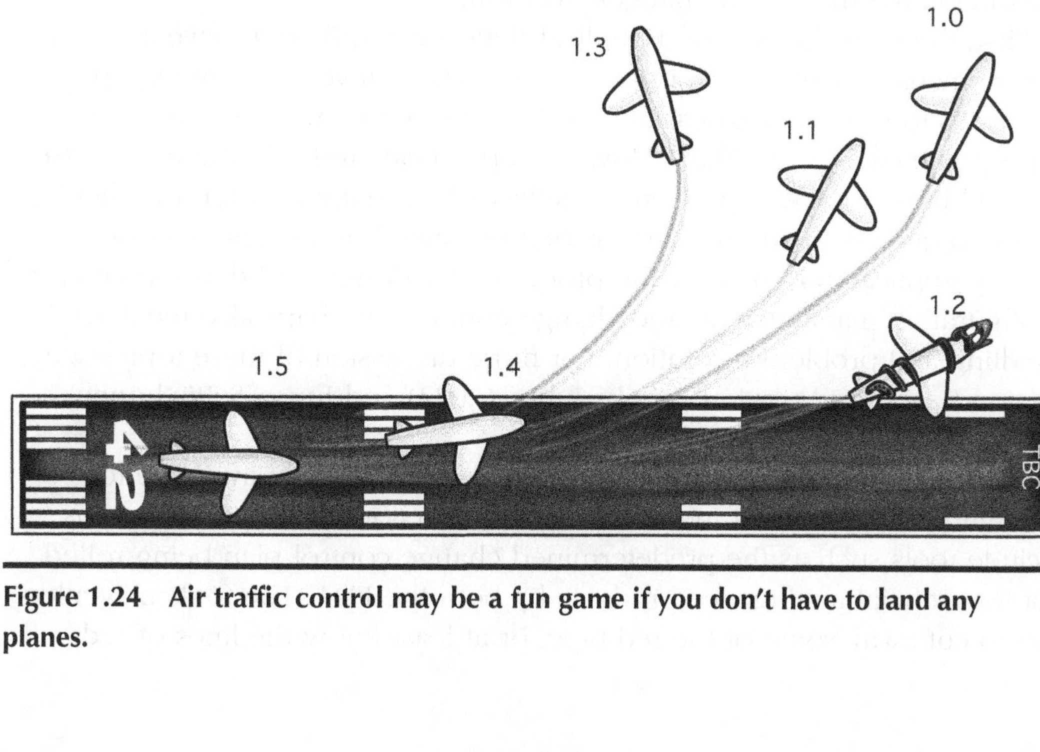

Figure 1.24 Air traffic control may be a fun game if you don't have to land any planes.

For work taking place over releases of the product, the answer becomes more complicated. In principle, the situation reverts to the previous lesson of when to rerun evaluations (see Section 1.22) but it will also be deeply impacted by your process for configuration management and change control (see Juuso & Pöyhönen 2024). In such circumstances, usability engineering activities might not be in the driver's seat as the overall process for addressing changes and the regulatory conformity of these changes might be the playbook you must now go by. Usability engineering should, of course, work within these processes to ensure that safety is maintained and, as a secondary consideration, also good usability is preserved.

The wish to push out quick updates to medical devices is something young software startups especially profess to. There is a certain level of built-in friction, initial inertia, and general sluggishness to releasing medical devices that feel foreign to the wider domain of software development out there. Pushing out updates and seeing what sticks, or what issues the users then face, is not a sound way of publishing medical devices. Elon Musk has rather famously said that if you are not breaking things, you are not moving fast enough. This from a guy who has taught cars to drive by themselves and put rockets and satellites into space – but he hasn't yet built a medical device, though. He is working on it, of course. Expecting this sort of mutagenic, or genetic-algorithm approach to pass in the medical-device field is a relatively quixotic endeavor. There are, however, mechanisms for simplifying the making of changes over time.

First, the manufacturer of a medical device is required to have a quality management system in place. This system involves drawing up processes to run the business and both develop and maintain products as appropriate (see Juuso 2022). How lean, practical, and safe these processes are, will have a great impact on the agility of the manufacturer and their ability to quickly push out versions of a product. The processes most directly implicated in this are the processes for design and development, configuration management, and change control – perhaps also feedback handling and problem resolution. For more discussion of these topics see Section 9, and our previous books Juuso (2022) and Juuso & Pöyhönen (2024).

Secondly, some special regulatory mechanisms may be available to streamline the release of incremental future versions. These mechanisms include tools such as the predetermined change control plan being rolled out for artificial intelligence products by both the FDA. The tools may allow you to cut away some of the red tape, or at least know the lines of red tape between which you are expected to work for each change, and thus better apply your resources. The longstanding practice of jointly agreeing on an appropriate configuration and change control scheme between the manufacturer and the notified body in Europe also has a similar effect for all types of medical-device products.

Thirdly, we would suggest that attending to the post-market surveillance of your medical devices is a great context for looking at how the usability of your devices might be called to evolve over time. By occasionally glancing up at the horizon you might have more time to appropriately respond to usability challenges and opportunities than you might have once some pressing issues have already emerged. This way you might also be able to take a calmer, more collected look at the usability of the implicated change, perhaps via formative methods.

1.24 Soft answers to hard requirements

Fluffy bunny feet is the punchline of an oddball joke played in the Hot Shots, *but it's also how the usability engineering standard and guidance occasionally read. Usability issues can have very real consequences for the users of medical devices, but the path to good usability appears to be paved more with good intentions than clear-cut, one-size-fits-all engineering activities. Occasionally the answers may be soft, but the underlying requirements stay firm as* Figure 1.25 *illustrates.*

We touched on this in a previous lesson discussing the amount of usability engineering considered to be sufficient (Section 1.20), but here we raise the stakes with one more seemingly heretic claim: it's not just the quantity but also the quality of usability engineering activities that are up for debate. Neither the standard nor its related guidance give us hard waypoints or comprehensive acceptance criteria for our activities, but both do want to be helpful.

The IEC 62366-1 standard stops short of giving us a readily executable model or a well-defined toolbox for usability engineering. The standard does give us a good, philosophical schema for usability engineering and enough nozzles to breathe life into the processes behind those expectations, but this still involves more blowing up of balloons than a two-year-old's birthday party. Similarly, the nonbinding guidance on usability engineering, IEC 62366-2, is at times written from a fairly aspirational point of view. The guidance paints a wistful picture of usability engineering as

Figure 1.25 Hard requirements deserve appropriate responses.

an activity with ample resources, deep pockets, numerous iterations, and an endless pool of experts and test subjects to draw upon in the name of creating the perfect device. The end of the rainbow would no doubt show us the perfect medical-device user interface, but the problem is that not many companies will have the stamina or the resources to get there.

In practice we need something a little bit more agile and reasonable than just the best of intentions – even if we do want to pay serious attention to the usability and safety of our devices. The standard and the guidance, IEC 62366-1 and IEC 62366-2, will help us get there, but not by reading every ‘should’ and ‘could’ as ‘must’, and not by endless repetition and reiteration.

The hard bottom line is that doing something about usability engineering is going to work better for you in regulatory audits than doing nothing at all. Just ticking a box may be enough to technically meet the requirements, but it may not be the smartest way forward for your device or organization. To reap the benefits usability engineering can yield you should aim higher than just perfunctory usability engineering, but you should also know what resources you are willing to expend to get there.

This book is written to address both the requirements and the optional guidance in place to help us meet the requirements. If the requirement has more exceptions than firm demands, or if the guidance appears to second-guess the standard itself, we will point this out. In the text between these covers we make use of must, should, and could as you'd expect in any engineering discipline – in agreement with the standard, but coming from a practical, results-driven way of thinking.

1.25 The guidance does not get to change the standard

Overlooked siblings have an uncanny way of turning into bad guys in more films than just Kevin Costner's Bodyguard, *but for now the good news is that the standard's chatty sibling, IEC 62366-2, does not get to make any demands. The meat of usability engineering for medical devices is in the standard as* Figure 1.26 *knows.*

Figure 1.26 The legendary video game was not called Super Luigi Brothers.

The IEC 62366-1 standard on usability engineering for medical devices is accompanied by an optional guidance document, IEC 62366-2, offering further nonbinding thinking on the use of the standard. Generally speaking, it is great to have guidance discussing the requirements to ensure our own understanding of the expectations is appropriate. That is true here as well as the guidance is built on a wealth of experience from usability engineering activities conducted over decades in both the US and Europe. However, this time we do feel that the guidance is quite a bit more verbose and more repetitive than we would like. We say this with some trembling as we fully expect some to want to roll us in tar and feathers for such a heretic remark, but there really is no getting around it: the present guidance is a nice-to-have document instead of a real must-read.

The IEC 62366-1 and IEC 62366-2 documents are like Mario and Luigi in the legendary Super Mario Bros video game by Nintendo. Mario is quite clearly the star and knows all the ins and outs of plumbing, while his green carbon-copy brother Luigi insists on tagging along. In the game, one is the headline star and the other the spare. This is mirrored in the two usability engineering documents, too. In the Nintendo-verse, Luigi found a new rewarding life as a ghostbuster of sorts traveling around haunted mansions. Perhaps the guidance too will find new life once it's revised.

In other words, if after reading the standard and this book you find yourself craving more you may look up the guidance. Work has already started behind the scenes to overhaul the present guidance as great many others too have voiced concerns similar to the ones we noted above. As part of the revision, the whole structure of the document is receiving attention, and the total of some 600 other comments received at the outset of the revision work indicate that a significant number of edits will be made to the guidance – but no changes whatsoever to the standard itself. At the end of the process, we will hopefully get a better, more linear, more succinct, and more practical (to use) guidance document, but for now Mario should really suffice. While on the topic of new thinking around usability engineering, some early work has also started on a guidance document for how usability engineering may be applied to artificial intelligence (this may become IEC 62366-3 in due course). This we feel is an important flavor to usability, and one whose footprint is only set to grow in the times ahead (see Section 11.2).

One final word of caution on the guidance document, both the present TR and the TS version set to follow it, is that the document occasionally overshoots the actual requirements of the standard. The intention here is noble: to explain best practices and solutions that make sense based on long experience. These breaches are, in theory, marked up as italicized text in the document, but this convention is easily missed as it is only mentioned in passing in the foreword of the document when discussing print types by saying: "Additional information about USABILITY ENGINEERING best practices: italic type". In practice, you should consider the whole of the guidance document as helpful discussion of what you might want to do to meet the requirements of the standard, whereas the actual requirements are to be found in the standard itself.

In this book we look at the requirements of the standard next in Section 2, then discuss the tools used in usability engineering in Section 3, and the resulting documentation in Section 4.

Chapter 2

Usability Engineering

Intuitive usability is a deceptively simple and dangerously alluring concept for the do-it-yourself enthusiast. It's tempting to think that as long as I put enough effort, attention, and time into polishing something it will surely become the best possible medical-device product for the world to use. Everyone sort of knows what good usability is, or so they think, but no-one can fully step into the shoes of another person and know exactly what each of them expects from the devices they will use in all the possible contexts they might use these in. This is where standards on medical-device usability engineering come into the picture to address what is inside and what is outside of the remit, as Figure 2.1 *illustrates.*

In Section 1.1, we were lightning quick to assert that usability for medical devices means first and foremost patient and user safety. Usability for medical devices does not mean pretty interfaces or shaving off seconds from clicking buttons. Not usually. In other words, in the world of medical devices, safety rides up front in first class and any notions of coolness, uniqueness, general visual appeal, and the like take a seat far back in economy if they even make the plane at all. As long as a medical device is safe to use, it doesn't really matter if it is also pretty or efficient to use. That is one part of why at medical-device trade fairs like the massive annual MEDICA fair in Düsseldorf, Germany, you still today occasionally see software that you think belongs in the early 1990s. Another part of this is that users are perhaps accustomed to these interfaces, and yet another that changing the interfaces would require time, money, and effort throughout the whole stack of the code, not just the interface perhaps.

DOI: 10.4324/9781003645702-2

Figure 2.1 Safety is the box for usability engineering, but usability also exists as a wider concept.

This is not necessarily the case for usability in general or for other, non-medical types of products. Outside of the admittedly very special context of medical devices, usability engineering is viewed as a discipline or a practice looking at how intuitive, fast, or otherwise pleasing products are to use. A non-medical product may be a complete nightmare to use for some of its users, but if it works great for most users most of the time it may be considered a success story. Cutting away minutes from the use of a software tool for each user may add up to great savings across a whole company, and thus it makes sense to look at averages and aggregates. In medical devices, however, individual nightmares may lead to adverse outcomes and lost patient lives. Here, looking at averages and aggregates is also of interest, but the outliers may quickly poison the whole value proposition of the tool. The two worlds of usability are not incompatible with each other – the drive to develop intuitive and efficient tools is naturally shared across both domains – but the key performance indicators (KPIs) are in all likelihood markedly different on the opposite sides of this line of demarcation.

Work in general usability engineering is guided by principles such as those introduced by the legendary usability guru, Jakob Nielsen, in his landmark 1994 book simply called "Usability engineering". Despite its advanced age the book is still used today. On the general usability front the objective is largely to polish the kinks out of products and speed up their use by meaningful factors of time. Six minutes here and two minutes there, multiplied by the total number of users, may add up to hundreds of thousands of dollars in either saved or wasted money. Hidden shortcuts may catapult power users to unlock even greater achievements. There is little doubt in similar benefits also being of interest to the users of medical devices, but in the tightly regulated domain of medical devices even a single adverse event or a lost life may bring the product to a grinding halt. The calculus thus becomes one of medical benefit and risk instead of purely the economy of scales or some wow factor. In a medical context time may, of course, matter greatly. After a cardiac arrest, for example, it is important that the automatic external defibrillator (AED) doesn't require a complicated boot-up sequence or hours of training to use. At the same time, a few minutes here and there matter much less in medical devices than the occasional lost life, and therein lies the reason why the term "usability engineering" can't just be loosely interpreted as meaning the same as in any other domain. We will discuss general usability in Section 2.6. For most of the rest of this book we will leave any performance sports apparel at the door, dress ourselves for the hospital operating room, and ground all our discussion in safety.

In this section, we will look at the actual requirements placed on usability engineering for medical-device user interfaces through the established IEC 62366-1 standard. Use of the standard is optional, but for tackling usability in the context of medical devices it is the smartest possible starting square. The following sections build on this picture by looking at the tools we can use in meeting the expectations (Section 3), and the documentation expected of that work (Section 4).

In its introduction the IEC 62366-1 standard, which should be considered as gospel on usability engineering for medical devices, makes the argument that medical devices are only becoming more commonplace in healthcare contexts. Doctors and nurses rely on devices to provide them with information on the patients, and treat patients in increasingly complex ways. Not only this, but also the patients themselves are becoming users of the devices who, without constant supervision, manipulate the devices to both observe and treat themselves. This much is briefly admitted by the introduction and a note to Clause 3.10 in the standard. What is not mentioned by the standard is that the devices, too, are increasing in complexity by evolving from syringes to automated agents, from simple what-you-see-is-what-you-get mechanical tools to automated systems perhaps even powered by artificial-intelligence brains. Such advanced medical devices may make it easier to engage with complex aspects of disease and disability through more intuitive concepts than was possible previously, but they may also jump up to more abstract levels of concepts which are less intuitive to comprehend even for professionals. The distinction between a white-box and a black-box device – i.e., something we can explain versus something we only think is working with the right outcomes – is also of interest here (see Section 7).

What the standard also doesn't mention is that the devices are at the same time becoming more fluid in their makeup, or more likely to evolve over their period of use. A software-based medical device might, for example, get more functionality over time and the functionality it had may also change as more evidence is churned up by the manufacturer or the scientific and clinical worlds. Keeping the user informed of what is happening under the hood is no longer a once-off event, but rather the user is to be kept informed of at least the significant changes taking place over time. At the same time, the manufacturer may also want to know that they are developing the device in the right direction, and want to hear from the users on this account also over the long term. These are topics for Sections 7 and 8, but here they serve to show that usability is a bigger task than just building the interface you yourself want here and now.

In medical devices, just about everything is viewed through a safety lens (see Section 1.1). The fundamental objective of risk management, as instructed by ISO 14971, is to achieve safety from unacceptable harm (see Juuso & Pöyhönen 2024). Similarly, the prime objective of usability engineering, as instructed by IEC 62366-1, is to identify and minimize use errors and thereby reduce use-associated risks. The overarching objective is thus to clad Schwarzenegger with as much padded gear and styrofoam as is needed to keep him safe on his journeys, regardless of how all this protective gear makes him look when he battles wizards, liquid-metal foes, and old age – unless, of course, our risk management activities indicate that he would have a higher survival rate wearing a swimsuit.

The scope of IEC 62366-1 thus lies squarely on the safety of a medical device. The standard instructs a process that is intended, and indeed presumed, to lead to devices that are acceptably safe to use, unless objective evidence is produced to the contrary (IEC 62366-1 Clause 1). A non-binding note (i.e., a recommendation, not a requirement) in the standard also implicates post-production information, i.e., post-market surveillance, in the continued monitoring of such evidence.

Interestingly, the IEC 62366-1 standard sees itself as addressing the analysis, specification, development, and evaluation of usability as this relates to the safety of medical devices, but only as far as normal use (including correct use and use errors) is concerned. In this the standard looks at the intended use defined for the medical device, or more broadly the use of the device in its intended use environment, as its roadmap to understanding normal use. The distinction made by the standard between an intended use and use in an intended environment begins to blur the border between normal use and abnormal use since both can take place in the same environment. We might look at the intended use as the focal point of our usability activities, but also broaden out from this to assess how the device is actually being used in the real world beyond our laboratory. As other standards and regulations point out, post-market information may also be helpful here over the long term.

The standard may help in the identification of risks regarding abnormal use, but it washes its hands of any assessment or mitigation of these risks. Furthermore, Clause 3.1 appears to absolve the manufacturer of the responsibility of mitigating such risks by defining abnormal use as a willful act (or an omission of an act) that violates normal use and is beyond reasonable UI-related risk control. Thus, technically, something which could be mitigated is not to be considered abnormal use even if it is not part of the intended, normal use. Examples given here of abnormal use include reckless use, sabotage, and deliberate disregard of information for safety. The standard does, however, point out in a non-binding note that also risk controls external to the user interface should be considered in the case of abnormal use. Finally, the decision to use a particular device in some clinical context is not a topic covered by the standard. Note that a similar disclaimer is also made by the industry-standard ISO 14971 risk management in instructing its application.

It is thus pertinent to consider the different types of use for medical devices envisaged by IEC 62366-1. These can be classified as follows:

- **Normal use** (Clause 3.9)
  This essentially corresponds to use as instructed by the Instructions for Use (IFU), or if this is not available, the generally accepted practice. This is explicitly stated to include routine inspection and adjustment by the user, as well as standby of the device. It is perhaps easy to equate normal use with the intended use of the device, but the standard is adamant here that normal use is a wider concept and, for example,

also includes the maintenance and transportation of the device. The two subclasses of normal use are seen as correct use (error-free use in accordance with the intended use) and use error (see Section 1.6).

- **Abnormal use** (Clause 3.1)
  This corresponds to any type of use which is not considered either correct use or a use error. In other words, a conscious or deliberate act (or omission of an act) in violation of normal use and which is beyond reasonable UI-related risk control efforts. See discussion above.

In saying that normal use comprises 'correct use' and 'use error', the standard appears to say that to err is human, but it is not correct. The standard also considers both 'use error' and 'abnormal use' as foreseeable misuse in the same sense that ISO 14971 defines this term. It then feels the need to wax poetic and forges a distinction between reasonably foreseeable use (everything it covers, including reasonably foreseeable misuse) and not reasonably foreseeable use (everything left outside of the standard). The rabbit hole here appears shallow, but full of quicksand and prone to misinterpretation if glanced at quickly.

The relationship between 'intended use' and 'correct use' appears obvious, but a nonbinding note to the definition of term 'normal use' (Clause 3.9) insists that these are not synonymous. According to the note, intended use is focused on the medical purpose of the device while correct use also considers non-medical use during, for example, maintenance and transport. We might also describe the difference here as the difference between an intended use (which is the root for all discussion of any medical device) and use in the intended environment (which the IEC 62366-1 standard wants to focus on), although this might needlessly brush aside some relevant environments other than the main use environment.

Figure 2.2 provides an overview of the different types of use addressed by IEC 62366-1. In the figure, normal use is shown on the left and abnormal use on the right. Enabling normal use is the goal for all design work while guarding against abnormal use is a topic for risk mitigation – and usability engineering may have something to say on both fronts.

The remainder of this section addresses what IEC 62366-1 has to say on the prerequisites, process, and deliverables of usability engineering. The standard itself is a brief dozen-plus-page sortie into methodically approaching some understanding of usability and the user. It is, nonetheless, an expected piece of reading on this topic.

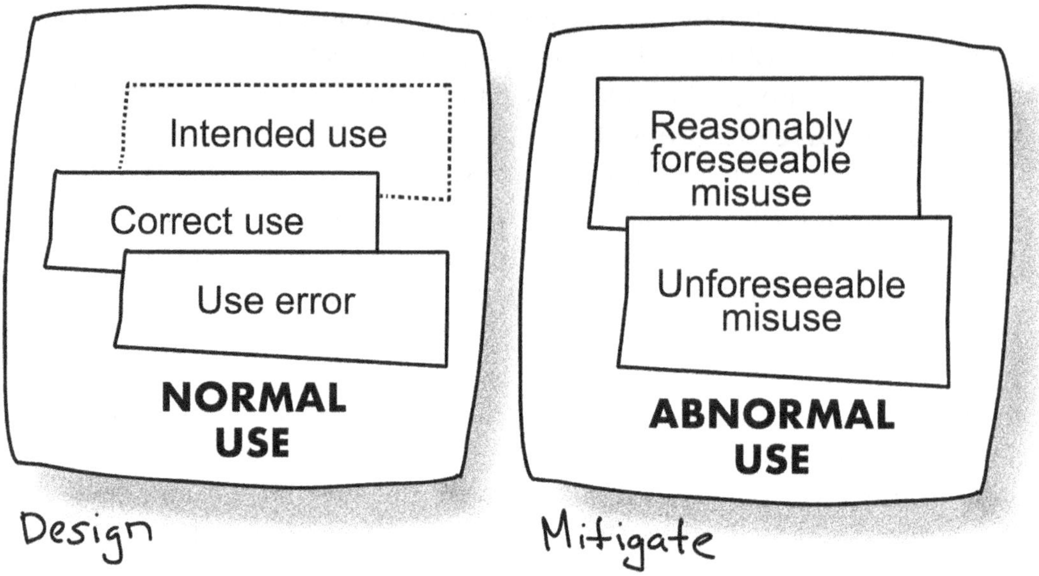

Figure 2.2 Types of use according to IEC 62366-1.

2.1 Prerequisites

The IEC 62366-1 standard builds a normative (i.e., binding) reference to ISO 14971 risk management. Thus, the standard is clear that it expects usability engineering efforts to lead to safer devices in the same way as ISO 14971 risk management. The reference is to the 2019 version of the standard, which is the latest revision and also the one expected by all current medical-device standards. Looking at how standards are evolving at the moment, the draft of IEC 62304 in particular, the normative reference to ISO 14971 may end up being dropped in a future version of IEC 62366-1 to facilitate a supposedly simpler adoption of the usability standard for devices other than medical devices, but even if this would occur we would argue that ISO 14971 is still the standard you should be looking to for risk management. Many customers, auditors, and regulators would still be oriented toward ISO 14971 – and moreover it is a smart, concise standard to follow on the topic.

The bibliography given at the end of the standard lists a further 49 publications as optional references to take into consideration when implementing the standard. These include publications such as the ISO 13485 standard on quality management systems, the IEC 60601-1 family of standards on the safety of electrical devices, and a range of ANSI/AAMI and FDA documents on, for example, human factors engineering.

The list also provides some interesting papers and books to refer to for more thinking on the topic of producing safe medical devices through usability engineering. The reading list is not provided as a further layer of requirements, only as helpful thinking on the topic.

2.2 Process

According to IEC 62366-1 a process for usability engineering is required to be established, documented, implemented, and maintained (Clause 4.1.1). The process must also be linked with any product realization (see Sections 1.10, 9.1) and risk management activities (see Sections 1.16, 9.2). With these assertions the standard drives a mountaineering anchor into the set of processes for any organization that finds that ISO 13485 Clause 7 on product realization applies to them.

The objective of the usability engineering process is to ensure the development of safe medical devices when these are interacted with as described in their accompanying documentation (see Section 3). This interaction is noted to refer to more than just the intended use of the device, and cover, for example, storage, transport, installation, operation, maintenance, repair, and disposal. To achieve these goals, the process wants to see usability engineering activities planned, performed, and documented by competent personnel. As in ISO 13485 quality management in general, personnel competence may be established based on appropriate education, training, skills, and experience.

IEC 62366-1 Clause 4.3 addresses the need to scale usability engineering activities based on the anticipated magnitude of the undertaken user interface development work. Factors to consider in choosing what methods and tools are appropriate may, for example, rest on the following:

- Complexity of the user interface
- Severity of user interface-related harms
- Complexity of the use specification
- Presence of UOUP components (see Section 1.18)
- Extent of modification (when modifying a medical device already subjected to usability engineering)

In practice, the process you define should address usability engineering with appropriate flexibility perhaps based on aspects such as those

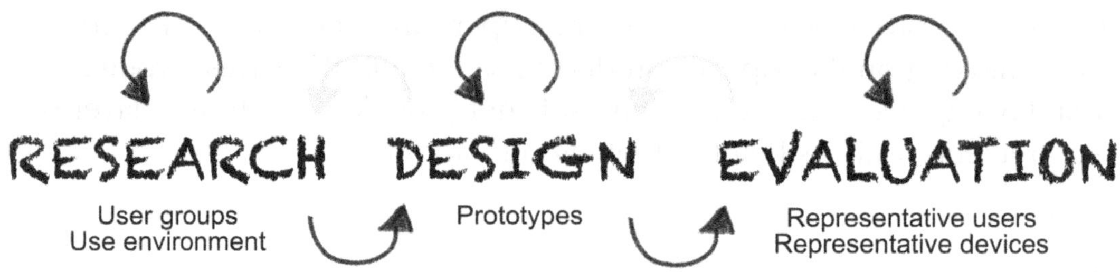

Figure 2.3 An idealized model of a usability process.

listed above. Later when assessing usability engineering records some documented rationale for the applicability of the above, as well as their impact on the chosen process, is typically expected here.

The informative guidance on the standard, IEC 62366-2, discusses a basic, idealized process for usability engineering. It sees the process as a potentially iterative path from research, design, and modelling to cycles of evaluation. Figure 2.3 illustrates this idealized model. In the figure we have fused modeling into the design phase as we see prototypes as a tool used to iterate the design rather than as something separate standing between design and evaluation. The figure also sketches the evolution of our understanding of user groups and use environments during early research phases to the utilization of representative users and devices in the final usability evaluations. Prototypes sit firmly in the middle of this path and while they could even be used in early research, they would not be permitted in the final summative usability evaluations unless they end up being the device itself. There is no right or wrong answer here as long as we preserve the basic plan-do-check-act cycle expected of all our activities (see Section 1.10).

The meat of the standard's take on usability engineering is discussed through Clauses 5.1 to 5.9 in IEC 62366-1. A further tenth subclause addresses the use of UOUP components which have a special, abbreviated pathway to conformity with the standard (see Section 2.5). These are discussed in the following.

2.3 Planning

A project plan for conducting usability engineering is generally recommended but not actually required by the standard (although read on). If such a plan is created, it may be integrated into other medical-device planning documents (e.g., software development plan) as is practical. The plan might build on the usability engineering process discussed above (see Section 2.2) by addressing, for example, the following topics:

- The goal of the planned usability engineering project (e.g., identifying use errors, improving safety, improving user satisfaction)
- The jurisdictions of interest to the project, and the regulatory requirements set by these jurisdictions on usability
- The skills and expertise required of the project team (including any subcontractors)
- The schedule, budget, and other resources allocated for the work
- The parts of user interaction to be addressed by the work (e.g., user interface controls, labels, instructions for use, quick guides, user support components)
- The use of optional formative evaluations and their timing
- The use of required summative evaluations and their timing
- The recruitment of test subjects for the evaluations (see Section 2.3.7)
- Development of usability engineering records
- Development of specific required conformity records

Note that a project plan may not strictly speaking be required, but a plan for user interface evaluation is required a little further down the road (see Section 2.3.7). Technically you could do all the research and design work and only prepare a plan once you are ready to move to the evaluation phase, but a firm grasp of the overall project and the resources needed in it is a smart goal from the very beginning of the project. The ISO 13485 standard on quality management also has expectations on planning (see Juuso 2022), and your usability planning should fit into this body of work. A project plan, therefore, probably should address at least some of the above topics. A standard operating procedure (SOP) instructing the usability engineering process will go a long way as a substitute for a plan, or as the necessary scaffolding to construct a plan, but in practice you will want to have a plan of some sort to also meet the basic plan-do-check-act cycle expectations (see Section 1.10).

The biggest benefit of developing a project plan for usability engineering may be to avoid slipping into last-minute usability work and then potentially having to either scuttle a launch if serious issues are detected or end up with a Mount Everest-sized dump of issues and aborted dreams under the carpet. If usability engineering is used to just provide a final go/no-go check on the medical device, this last-minute check may achieve the goal of ensuring the safety of the device or we may just be closing our eyes to some problems in order to push the device through. In any case, a last-minute activity performed under pressure will not be conducive to analyzing the usability of the device or providing insights on how it could be improved. Detecting the issues earlier – by even using paper prototypes at the beginning of development work – we will be improving usability and perhaps even avoiding an ulcer or two.

Whether or not you develop a specific project plan, you should ensure that the usability engineering team has the appropriate expertise to perform their jobs. This is stated in the non-binding guidance to the usability engineering standard, IEC 62366-2, but it is also something ISO 13485 quality management expects to see checked and documented. In other words, it is not enough to have representative users taking part in the summative usability evaluations, as you should also make sure that the experts evaluating the results of the tests pass muster. According to the guidance, your whole team does not have to consist of usability specialists alone but can also include, for example, technical writers, trainers, marketing specialists, engineers, and clinicians. In fact, the guidance takes a rather relaxed view of who qualifies as a 'usability specialist': the necessary skillset may come from education or experience, from formal training or self-education. What matters to the standard is that the chosen specialists are in tune with analyzing the usability of the user interfaces in question, and collectively have appropriate medical-device domain knowledge. You thus have great freedom in building your team, but you should know why the team meets your expectations and be prepared to defend your choices during conformity assessment as well.

The guidance also discusses the need to ensure appropriate resources for the usability engineering project. This too harks back to ISO 13485 which places similar expectations on the manufacturer's top management to provide adequate resources for quality management in general. If usability engineering is made a part of the processes of the quality management system, it too falls under the same weighty requirement. In practice, you are fairly free to assess the level of usability engineering efforts needed for your products, but you should ensure that you provide adequate resources to then perform that work. The guidance, too, points out that usability engineering shouldn't be a siloed effort, but instead integrate into other quality management system processes (see Section 9).

2.3.1 Preparing a use specification

The first step in usability engineering is to begin drafting a use specification based on everything we know about the device and its use. Here, IEC 62366-1 Clause 5.1 formally requires the manufacturer to prepare a use specification, which was earlier defined as a summary of the key characteristics related to the context of use for the medical device (see Section 1.11). This must now include the following:

- Intended medical indication
- Intended patient population (e.g., age group, weight range, and health)
- Intended body part or tissue type interacted with
- Intended use environment (e.g., sterility, reuse, home use, protective gear)
- Operating principle (e.g., laser, stainless steel blade, intravenous tube)

Note that whereas the original term definition in Clause 3.23 (see Section 1.11) presented a few of these items via non-binding notes the tone here in Clause 5.1 is no longer optional. You must have these elements within or linked to your use specification in some clearly stated manner. While the clause doesn't mention user profiles by name, a subset of these is in essence what is referred to by intended patient populations. You may have other user groups besides patients, of course, but developing user profiles here will help you in the next steps, and it is implicitly expected to take place by Clause 5.3 later.

The above list does not include the actual intended use of the medical device, but it does approach that description via many of the other aspects expected to be considered here. Again, a non-binding note to the clause points out that some jurisdiction may even expect a summary of this specification to in fact be the statement of the intended use. Nonbinding guidance suggests that sometimes an early use specification might indeed be a draft of the intended use. In any case, the use specification and the intended use should be compatible with each other.

In practice, Clause 5.1 expects you to understand the use of the medical device – not just the user interface of that device – and describe that in a practical set of documentation. You might pool all this information into a single document or produce a brief dossier pulling in the relevant information. The first approach may be practical for a small, one-of-a-kind project. For larger projects or projects building on a portfolio of existing devices it may be more sensible to refer to the appropriate documentation providing in-depth description of each aspect. The standard does not limit your choices here, but you must be able to show you have considered each aspect and that locating the constituent documentation is not an exercise in treasure hunting. Watching Nicolas Cage traverse his way through the colonial clues in the National Treasure films is entertaining, just like watching Tom Hanks stitch his unique path through American history in Forest Gump, but your auditors won't sit for such screenings. You might also lose the red thread yourself over the years.

For UOUP no exceptions to Clause 5.1 are made by annex C of IEC 62366-1, and thus everything that was said above applies. This is the only part of Clause 5 that is retained for all use related to UOUP components (including any changes related to them). For more discussion on UOUP see Section 2.5.

2.3.2 Identifying safety-related user interface characteristics

After we know what sort of a device we are creating, who will use it, and in what environment, it's time to identify any safety issues these may implicate. Accordingly, IEC 62366-1 Clause 5.2 tasks the manufacturer with identifying those characteristics of the user interface that could be related to the safety of the medical device. This activity is to link up with what is expected by ISO 14971 risk management, but identification of such characteristics related to risks might begin from either standard. Limiting our gaze to usability engineering will not be enough to identify all possible risks regarding the safety of the medical device, but at the same time not considering usability adequately may lead to some risks being missed.

Clause 5.2 insists that this identification of risks considers so called 'primary operating functions' (see Section 1.5) that are suggested by any applicable medical device safety standard. The 2020 amendment to IEC 62366-1 changed the wording slightly here, but the intention remains the same: if applicable medical-device safety standards exist, and if those provide primary operating functions, those are to be considered here. The notion here makes all the sense in the world: if someone already pointed out, while writing a relevant standard, that jumping off a skyscraper without a parachute is dangerous you should not claim that as a surprise when you forget to pack one for your first-ever base jump. If another standard points out that jumping with an unchecked parachute is dangerous for base jumping, you should also heed that caution. The early parachute-enthusiast Franz Reichelt (1878-1912) would perhaps have appreciated such standards in his day before jumping to his death from the newly constructed Eiffel Tower in Paris, France, but he also failed to do appropriate risk management or heed the warnings from his earlier formative experiments from lower heights and those using dummies. The usability engineering standard would not be happy with you for doing it either.

The total set of primary operating functions may thus be a combination of several standards and your own analysis of the situation. In the medical-device world it is often the standards of the IEC 60601 family of standards that should be consulted for such known facts. Let's consider an example of a pulse oximeter being used in a home healthcare setting. Here the set of pre-existing knowledge on primary operating features will include at least the following:

- The IEC 60601-1 base standard on General requirements for basic safety and essential performance
- The IEC 60601-1-11 standard on requirements for medical devices used in the home healthcare environment
- The IEC 60601-2-61 standard on particular requirements for basic safety and essential performance of pulse oximeter equipment

Together this set of standards will, for example, point out that a typical hospital environment and a home setting will differ greatly, and in the latter, you should expect to also get some lay people as your users instead of only trained medical personnel. The standards provide a long array of potential safety factors for you to analyze.

The definition of a 'primary operating function', or POF for short, is not always intuitive to grasp when you first encounter the term. As we pointed out in Section 1.5, we might think of these as actions related to your device which are critical for the user to get right lest some harm might occur. Examples of potential POFs might thus include:

- Setting the alarm thresholds on a device
- Acknowledging or turning off alarm signals
- Switching between user modes (e.g., normal operation v. service mode)
- Activating some display or scale for measurements
- Securing a good contact for the pulse oximeter probe

Many of the above examples are adapted from the IEC 60601-2-61 standard on pulse oximeters. Similarly, the IEC 60601-1-11 standard on the homecare context might want you to look at the usability of accompanying documents for a layperson or the practicality of the instructed processes in a home setting.

However, typically the following would not be considered primary operating features:

- Turning a device on
- Moving a device over to its use site
- Mundane cleaning procedures (excluding sterilization)
- Mundane maintenance procedures
- Connecting additional devices to a system according to manufacturer instructions
- Trivial battery replacement (e.g., household AA batteries)

Note that in the above we said "typically". If your own risk management concludes that some risks are related to the device being switched on appropriately, for example, you should include that function among your primary operating functions. Remember for example the ventilator with an on/off toggle we discussed in Section 1.9, in that case switching the device on may have had implications for risk management but accidentally turning it off definitely should have featured in the risk matrix. Almost nothing about developing medical devices is as easy as "just add water" with some recipe you might be following. Instead, you must consider all the relevant factors and act appropriately – and following standards will help you consider all such factors.

Perhaps all the factors pointed out by the standards are relevant to your device, or perhaps only a handful truly are. The point here is to get you to think about all the foreseeable factors and figure out how they may influence your design and risk management. It will always be up to you to identify the relevant interactions, their sequences, and the functions used in them, but standards will point out some of the typical issues. Based on your initial analysis you will then pinpoint those functions which may be directly related to the safety of the device.

The level of detail to go into while defining interactions may also take some thought as a level that is too abstract won't be conducive to identifying all the right functions. Consider for example the use case of cleaning medical devices. This activity may be defined on a very high level, perhaps as just that, or by diving deeper to observe that cleaning devices actually calls for, for example, moving devices, preparing them for cleaning, choosing cleaning agents, wearing protective gear, applying a cleaning agent, letting the cleaning agent work its magic, removing the cleaning agent, and returning the device to use after appropriate checks. On the other hand, defining an activity in too much detail, or too many irrelevant steps, may mask the very insight we are after.

At the end of the day, the primary operating functions you identified will act as inputs for your design and risk management activities. Some identified functions will deserve much more attention in your usability engineering activities, for example in the formative and summative evaluations you plan to conduct.

Finally, building on both the use specification and the now identified user interface characteristics, the manufacturer is tasked with identifying any related use errors. As always, all results of this identification work are to be recorded in the usability engineering file. The wording in the 2020 amendment to IEC 62366-1 implies that in addition to characteristics related to safety we may also want to record potential use errors here.

For UOUP (see Section 2.5) Clause 5.2 is excluded in its entirety by annex C of IEC 62366-1. None of what was said above thus applies to UOUP components. Note that changes made to the UOUP may strip the changed parts of any such exemptions.

2.3.3 Identifying hazards and hazardous situations

The terms "hazard" and "hazardous situation" are terms defined by the ISO 14971 standard on risk management (see Juuso & Pöyhönen 2024). For our discussion here it is sufficient to think of a hazard as the source of danger and a hazardous situation as the context we are exposed to that danger. In terms of the Michael Bay film "Armageddon", a hazard would be the giant asteroid heading for Earth, and a hazardous situation perhaps the decision to land and drill on the asteroid – although this being an action-adventure film there are certainly plenty of both hazards and hazardous situations in the storyline. Some might also argue that the greatest hazard for human viewers of the film is the nauseating animal cracker Ben Affleck rides towards the navel region of Liv Tyler.

Be that as it may, IEC 62366-1 Clause 5.3 charges the manufacturer with identifying both known and foreseeable hazards and hazardous situations which patients, users and other people might find themselves faced with.

Annex B of the standard provides half a dozen examples of harms and associated risks related to usability. The list includes such examples as triggering a radiation dosage by accidentally pressing an unguarded button, displacing a sharp needle, and relying on a frozen or uncalibrated display. The list is nowhere near comprehensive, as indeed it could not be, but perhaps serves as food for thought.

The IEC 62366-1 standard expects the manufacturer to identify hazards and hazardous situations while also considering the following:

- The use specification (including user profiles)
- Any identified use errors
- Any information on hazards and hazardous situations regarding similar existing medical-device user interfaces

A non-binding note suggests that even abnormal use conditions could be identified through this work. In other words, even risks that are not related to the user interface, are outside of the use specification, and can't be controlled in the context of the user interface (including information for safety), might be at least discussed via abnormal use here. Note though that this in no way means that such risks would be automatically exempt from handling as part of the overall ISO 14971 risk management of the medical device. The hazards and hazardous situations conjured up here may turn out not to be relevant in the context of the user interface, but if they are relevant for the medical device itself, they must be reflected in the overall risk management of the device. Once again, all the results of this identification work are to be included in the usability engineering file.

For UOUP, Clause 5.3 is excluded in its entirety by annex C of IEC 62366-1. None of what was said above thus applies to UOUP components, except when changes are concerned. Note also that a requirement to review post-production and risk management information is set for the use of UOUP (see Section 2.5).

2.3.4 Identifying hazard-related use scenarios

Previously we identified hazards and hazardous situations related to our medical device and its user interface. Now IEC 62366-1 Clause 5.4 instructs us, as the manufacturer, to identify and describe the reasonably foreseeable hazard-related use scenarios associated with these hazards and hazardous situations. The standard defined the term "hazard-related use scenario" in its Clause 3.8

rather unsurprisingly as a use scenario that could lead to a hazardous situation or harm. In that clause such scenarios were noted to perhaps be linked with a use error, which might also then be implicated in a failure of the device.

In this clause we are to identify such use scenarios and describe the tasks (including their sequences) leading to the harm and the severity of that harm. Interestingly the severity of harm is not given any scale here, the term severity is not given a definition in the terms section of the standard, and the example scenarios given in annex B of the standard do not explicitly mention severity. In other words, we might here have ambiguous textual descriptions, such as "ELE" for an extinction-level event as made popular by Armageddon's rival Deep Impact. If, however, we consider the wider topic of risk management for our medical device we will likely end up defining some sort of scale to use here, for example, numeric scales from 1 to 5 or 0–100%. It is in fact the ISO 14971 standard on risk management that will guide us in defining severity in a standardized and defensible way (see Juuso 2022 for medical devices in general, and Juuso & Pöyhönen 2024 for software). After all, there is little point in defining risk severity on a different scale for usability-related risks and for all other risks.

In practice, the definition of hazard-related use scenarios should be mapped to the overall list of identified hazards and hazardous situations. Each scenario would thus have a severity (and a probability) defined for each associated risk as per ISO 14971, and a textual description of the tasks involved to realize the risk. The form and format of this description is up to the manufacturer, but might, for example, take the form shown in Table 2.1. The table also shows a possible user interface-related control for the risk.

Table 2.1 Example use scenario

Hazard	Explosive charges fail to detonate
Hazardous situation	An extinction-level asteroid is heading for Earth and needs to be blown up before it kills everyone. A team of oilmen fly to the asteroid to drill holes and plant explosives.
Use scenario	*Remote controllers require batteries.* Bruce forgot he took out the batteries to protect against accidental detonation. He now presses the button while sitting all alone on the asteroid, and nothing happens. Asteroid is not destroyed and remains on course towards Earth.
Harm	All life on Earth is killed. Severity: ELE, i.e., extinction-level event.
UI risk control	UI displays a power on -indicator so that it is easy to see the remote controller is ready to operate. This is safety by design.

The above example is astronomically off-topic for medical devices, but it does illustrate how the IEC 612366-1 standard expects us to define hazards, hazardous situations, and the use scenarios related to these. The standard does not in fact give us a wealth of guidance on how to write use scenarios. How ever you end up writing your scenarios, ensure that the content and form aligns well with your overall ISO 14971 risk management.

For UOUP (see Section 2.5) Clause 5.4 is excluded in its entirety by annex C of IEC 62366-1. None of what was said above thus applies to UOUP components. Note that changes made to the UOUP strip the changed parts of any such exemption.

2.3.5 Selecting use scenarios for summative evaluation

IEC 62366-1 Clause 5.5 charges the manufacturer with selecting those hazard-related use scenarios that are to be included in the summative evaluation (see Section 1.13). The selection here is to either include all the defined scenarios or a subset of scenarios based on the severity of the harm caused by use error. Both the standard and its 2020 amendment go on in their own ways to say that the selection scheme may also depend on the circumstances of the medical device and the manufacturer. In any case, description of the selection scheme, the rationale for its use, and the results of its application must be recorded.

In the Star Trek canon, a 'Kobayashi Maru' is a Starfleet training exercise that is intended to be impossible to win without also losing in another respect at the same time. Captain Kirk is naturally the only man in history to has successfully done the impossible, but even he did so by hacking the exercise. Cherry-picking the scenarios in your summative assessment to pass the evaluation is naturally not the goal here, but neither is testing for all of the Kobayashi Marus outside of your intended use that you could invent. Be practical here and use the occasion to ensure the tasks you intend to be accomplished with your device are safe to accomplish. The goal here is to ensure no usability-related issues stand in the way of the safe use of your device – as long as your usability activities capture this you may feel like you have accomplished the assignment.

For UOUP (see Section 2.5) Clause 5.5 is excluded in its entirety by annex C of IEC 62366-1. None of what was said above thus applies to UOUP components. Note that changes made to the UOUP, strip the changed parts of any such exemption.

2.3.6 Develop a user interface specification

IEC 62366-1 Clause 5.6 wants to see a 'user interface specification' established. This must take into consideration the previously developed a) use specification, b) identified use errors, and c) hazard-related use scenarios. The specification must include the following:

- **Testable technical requirements**
  The specification must include testable technical requirements for the user interface, including any required risk controls. These may, for example, include display color, capacitive versus resistive touch, and number or placement of controls.
- **Indication of whether accompanying documentation is required**
  A simple yes or no would satisfy the requirement of the standard here, but it may be worthwhile to consider why such documentation might be needed or not needed.
- **Indication whether device-specific training is required**
  Again, a simple yes or no would satisfy the requirement of the standard here, but it may be worth your while to consider why such training might be needed or not needed.

A non-binding annex to the standard, A5, suggests that the user interface specification contain acceptance criteria for the usability of the user interface. No such requirement is made by the standard itself, although the call to define requirements may be interpreted as designating the finish line to pass with the developed user interface. The annex even suggests factors to consider in defining the criteria. These include the state of the technology, experience with a similar device, and post-production surveillance. As always, the specification must be accessible via the usability engineering file, but it may be a self-contained specification or integrated into other appropriate specifications.

For UOUP (see Section 2.5) Clause 5.6 is excluded in its entirety by annex C of IEC 62366-1. None of what was said above thus applies to UOUP components. Note that changes made to the UOUP, strip the changed parts of any such exemption.

2.3.7 Develop a user interface evaluation plan

IEC 62366-1 Clause 5.7.1 asks the manufacturer to plan how the testing of their interface will be performed. As always in medical devices, the plan must be kept up to date. This doesn't necessarily mean updating the plan to log progress made according to the plan but instead updating the plan when your plans change (see Juuso 2022). Here we assume you have some other means of tracking and recording progress along the plan, such as the project management tool of your choice.

The plan should describe the objectives and methods for any planned formative and summative evaluations. In case usability tests are used, it should also address the following:

- The user groups intended to take part. In a pair of helpful non-binding notes the standard suggests that the manufacturer's own clinical personnel could be used for formative evaluation, while a panel of external, practicing healthcare professionals might be used for the summative evaluation. The binding text of the standard does allow for multiple user profiles to be combined into user groups (see Section 1.3) for this testing.
- Test environment and other relevant conditions as per the use specification (see Section 1.11).
- Indication whether accompanying documentation is provided during the test.
- Indication whether training is provided prior to the test, as well as the minimum interval time between training and testing.

The last two items on the list are implicitly presented as simple yes or no -choices by the standard, but we would be very weary of not providing further detail on exactly what material or training was made available. The scientific process relies on the repeatability and reproducibility of experiments, which is to say that to omit sufficient information here would court suspicion later.

The standard allows for both quantitative and qualitative methods of evaluation and appreciates both simulated and actual use environments. As always, the plan must be accessible via the usability engineering file, but it may be a self-contained plan or integrated into other appropriate plans. The clause addresses planning for both formative and summative evaluations, in part for both at the same time, and in part focusing on one form of evaluation at a time. Table 2.2 provides an overview of the planning requirements for each type of evaluation as set in the subclauses of clause 5.7.

Table 2.2 Requirements for planning formative and summative evaluation

	Formative	*Summative*
Aim	Anything the manufacturer deems worthwhile. Evaluation can be reperformed until the manufacturer is satisfied.	Provide objective evidence on the safety of the device (including the mitigation of usability risks).
Address all hazard-related use scenarios (as per the selection criteria defined in Section 2.3.5)	Not required.	Required.
Definition of correct use for each hazard-related use scenario	Not required.	Required.
Evaluation objectives and methods	Required.	Required, including rationale for the methods creating objective evidence. Also, the data collection method to be used during usability tests to support analysis of observed user errors and use difficulties must be specified.
Description of user groups	Required. May be performed using the manufacturer's own personnel.	Required. Personnel directly responsible for the design should not be used (see Section 2.4.2). A panel of representative real users is suggested. Justification is required for a) how the test participants match with representative users, and b) how user groups were allocated. Note that in the latter the number of test participants should be simple to retrieve and not obfuscated by the group allocation.

(*continued*)

Table 2.2 (Continued)

	Formative	*Summative*
Description of the test environment and other relevant conditions	Required.	Required. Justification is required for how the test environment and use conditions match with the intended use environment.
Identification of the part of the user interface addressed	Recommended.	Required.
Determine if information for safety is perceivable, understandable, and supports correct use of the medical device	Not explicitly required.	Required, where applicable.
Definition of any required accompanying documentation	Required (when usability tests are involved).	Required.
Definition of any required training	Required (when usability tests are involved).	Required. Remarks on a wait time between training and testing may be appropriate.
Acceptance criteria	Not required. Formative evaluation is performed until the manufacturer feels confident enough to proceed to summative evaluation.	Criteria are no longer required after the 2020-amendment, but assessment is required at the end covering a) whether use errors occurred, b) use difficulties occurred, or c) task were completed successfully.

It is noteworthy that the requirement for acceptance criteria changed between the 2015 and the 2020 versions of the standard. Previously, a non-binding annex, A5.7.3e, went beyond its mandate to suggest formal acceptance criteria were required for summative testing. The criteria were expected to specify when a task associated with a hazard-related use scenario could be judged as successfully completed by the user, and this was to be in line with the manufacturer's policy for setting the ISO 14971 risk acceptability criteria. This shotgun demand was not made by the body of the standard, it was only presented in a non-binding annex A, but now it is no more. Annex A5.9, however, continues to suggest that the summative evaluation plan is a good location for such criteria to be defined.

The requirement for defining acceptance criteria for summative evaluation is thus gone, but in its place is a newly forged hard requirement (see clause 5.9) for the manufacturer to assess if use error occurred, and if users had any difficulties in completing the tasks associated with hazard-related use scenarios. This notably expands the scope of study from a situation where something went wrong to situations where something could have gone wrong (i.e., close calls), or where users experienced momentary difficulties (i.e., use difficulties). In other words, acceptance criteria were perhaps already previously hanging on to the standard by their fingertips, but now someone stomped on those fingers.

Regardless of the now assassinated acceptance criteria, you may still be faced with questions on when a particular usability requirement should be judged as met or when an activity is complete. In this case, and in the absence of specific criteria for summative evaluations, you could refer to your product requirements and, in particular, the product-specific risk management file you have developed. The ultimate goal of summative evaluations is after all to evaluate the safe use of your device in accordance with your risk management activities. Coming up with new acceptability criteria for usability-related risks (which is what we would quickly be doing here), while using some other criteria for other types of risks, is probably not useful or even beneficial. Knowing what a completed summative evaluation should look like, on the other hand, is a matter to address via planning and goal setting, not just a feeling after a certain amount of time has passed. We would, however, just refer to this as planning and reporting following the standard PDCA model (see Section 1.10), and not confuse terms by labeling this as acceptance criteria.

Finally, note that the IEC 62366-2 guidance document also discusses something called a 'usability evaluation plan', which it sees as a part of the verification and validation work in product realization. The standard does not discuss any such document, so our recommendation would be to stick with the standard and not scatter our plans needlessly. In the name of not confusing matters we would mercilessly hit away at this faux mountaineering anchor until it disappeared into the clouds below. To do so is not to advocate for free climbing, but to focus on the actual requirements of usability engineering.

For UOUP (see Section 2.5) Clause 5.7 is excluded in its entirety by annex C of IEC 62366-1. None of what was said above thus applies to

UOUP components. Note that changes made to the UOUP, strip the changed parts of any such exemption.

Formative evaluation

Formative evaluation is the work horse of the standard. In Clause 5.7.2, the standard promotes the use of formative evaluation as early on as is possible during the design and development process of the device as this can shape the process by offering valuable information. The objective of the evaluation may be explorative, and, for example, look at how well the elements of the user interface are recognizable, understandable, and operable by the user. For this reason, and as noted by the non-binding A5.7.2 annex to the standard, formative evaluation usually does not have formal acceptance criteria. In other words, formative evaluation is both undertaken and perhaps repeated as seen to be beneficial by the manufacturer before proceeding to summative evaluation. It is technically possible to forgo formative testing completely and only perform summative evaluation. This may be akin to setting out on a voyage across the seas to discover new lands without any maps – if you still end up where you wanted you may have providence to give thanks even if your destination was not Providence, Rhode Island.

Formative evaluation may address the whole user interface at once or parts of the interface at a time, as is practical. In contrast to the later summative evaluation, which is required to address hazard-related use scenarios, the scenarios may or may not be referenced here.

Summative evaluation

Summative evaluation is the crown on the usability engineering work, and as such must as a body address the entire user interface to be developed. The objective of summative evaluation is to obtain objective evidence that the user interface can be used safely (see Section 1.13). In contrast to formative evaluation, summative evaluation must address each selected hazard-related use scenario. A non-binding note further suggests that the finalization of the plan for summative evaluation may have to wait for the formative evaluation to be completed. For this reason, it may be practical to separate the two plans, but that is not required by the standard.

The non-binding A5.7.3 annex to the standard notes that accompanying documentation should now be made available to the test subjects to facilitate realistic test conditions. Similarly, any training specified as required is to now be made available and you may also want to consider what is an appropriate interval between the training and the subsequent testing to allow for memory decay to take place. After all, total recall is something we knew to expect of Arnold Schwarzenegger in that sci-fi caper by the same name, but it is not something we can expect of our real users over the long term. The annex is adamant that if training is a risk control, we must ensure its appropriate and timely provision. The silver lining here is that if we do provide training and required documentation in a realistic manner here, and to the extent you deemed necessary, we can also assess their effectiveness as risk controls based on the summative evaluation conducted.

Here too the body of the evaluation work may consist of a single evaluation or multiple evaluations as is practical. In particular, a non-binding note to Clause 5.7.3 suggests that summative evaluation for safety may be completed before a summative evaluation of the user interface in general is begun. The two rounds of evaluation may also involve different users and different usability tests. The standard makes a somewhat oddly placed note here that information for safety may require different methods than the other parts of the user interface. The same is not said for formative evaluation, where information for safety is thus not implied to be a part of the picture. Perhaps this is defensible as you will want representative users to make sure the information you provide makes sense to them, and not just to those users you might have on your payroll. Nonetheless you will probably want to make sure anything you send out for testing by real-world users first passes your own evaluation, be that the design of the device or the related information you might supply.

2.4 Conducting evaluation

Now that we have a plan it's time to execute on that plan. In the following we'll discuss both formative and summative evaluations, one at a time. Think of these two steps as showing a rough cut of your film to a friend in the editing room (formative) and as the ultimate audience screening of the finished film (summative). With both steps the aim is to predict how well your finished product will sit with its audiences, but with the latter the stakes are considerably higher.

2.4.1 Perform formative evaluation during development

Did we already say that the IEC 62366-1 expects you to perform formative usability evaluation during your design and development work? We did, didn't we. The standard drives this point hard and mentions the expectation in several different clauses. Remember this when defining your ISO 13485 product realization processes (see Juuso 2022, Juuso & Pöyhönen 2024).

In the standard it is Clause 5.8 that gets the honor of waving the pace car off the track and letting you execute on the plans you have made. Here you, as the manufacturer, must respect the user interface specification you made earlier (see Section 2.3.6), and remember what your formative evaluation plan thought would happen during the design and development work. You must design and implement the interface you specified, including any accompanying documentation and training capability you may have promised there – or if that is not achievable go back to the drawing board as is appropriate. The standard also doubles down on expecting you to make use of formative evaluation and other usability engineering methods and techniques as appropriate during the work.

The non-binding annex to the standards, A5.8, suggests that you may want a multidisciplinary team working on designing your interface to consider as many relevant points of view as is possible. This team might, for example, include engineers, UI specialists, cognitive psychologists, usability engineers, and training specialists. The one thing the standard dreads the most is an interface designed by just one person applying their own brand of common sense, or no sense at all.

If you discover new use errors, hazards, hazardous situations or hazard-related use scenarios you must look back to update or redo the work instructed in Clause 5 (see Sections 2.3.1 to 2.5) as is appropriate. This will also apply to design changes made in the device.

In terms of the training you have deemed required for the safe use of the device, you are instructed to design and implement the required training capability by at least making training and/or training materials available. The training may also be made available to responsible organizations (see Section 1.15) who then train the users under them. Note also that it is not enough to have the requisite training capability at launch, but you must also plan for the expected service life of the medical device.

For UOUP (see Section 2.5) Clause 5.8 is excluded in its entirety by annex C of IEC 62366-1. None of what was said above thus applies to UOUP components. Note that changes made to the UOUP strip the changed parts of any such exemption.

2.4.2 Perform summative evaluation at the end

After it is all built, packaged along with any accompanying information, and almost ready to ship to its users you're instructed to pass one more crucial gate: a summative evaluation. As explained above, this might be one gate or several gates in sequence. What matters at the end is that you have obtained objective evidence on the safety of the user interface (and by extension the medical device itself), evaluated it, and hopefully received a clean bill of health to proceed with your world domination plans.

In IEC 62366-1, it is Clause 5.9 that carries the punch regarding performing summative evaluations. A lot of the details were already figured out while making the corresponding plan in Section 2.3.7. Here you are again required to cover all of the hazard-related use scenarios you earmarked earlier, and perform evaluation using the final or production-equivalent user interface you developed. You must now use representative users as the test subjects, and – according to the non-binding annex A5.9 – these should not include personnel who were directly responsible for designing the user interface. Note that you may also use data obtained from summative evaluations of products with an equivalent user interface if you provide a sound rationale for its applicability.

The summative evaluation data must be analyzed with respect to all the use errors and use difficulties encountered. As nearly all summative assessments dredge up use errors, the question here is not did we get any use errors, but what should we do about each use error we got. Thus, the real questions to answer here are:

1) What are the potential consequences of each occurred use error? If the use errors can lead to a hazardous situation, what is – based on, for example, observation of user performance and any remarks on that performance they make – the root cause behind the error or difficulty?
2) Did any new use errors, hazards, hazardous situations, or hazard-related use scenarios emerge during the evaluation?
3) Is further work on improving the safety of the user interface necessary or practicable?

The answer to the first question is what will probably occupy our time the most here. The notes, observations, and user comments from testing will help you arrive at the true cause of the error. You may also notice that multiple use errors are related, or that an occurred use error suggests also another hazard-related use scenario may be likely – and the severity of the harm there is perhaps greater. You may also discover that your designed risk controls are ineffective and need revision.

The answer to the second question may be more straightforward, but it will also plot our point along the quest to finish the user interface and, quite possibly, the medical device itself. If the answer here is yes, we must revisit the user interface specification (see Section 2.3.6) and repeat any affected work. If, on the other hand, the answer to the question is no, we must still assess whether we ourselves are happy moving forward. This is done by answering the third question.

In assessing the answer to the third question, a yes-answer naturally takes us back to revising the user interface specification (see Section 2.3.6). If the answer to question 3 is no, we must a) document our rationale, b) identify the relevant usability engineering data needed in the determination of any residual risk for the use of the interface, and c) evaluate the residual risk in accordance with ISO 14971. Importantly, also note that the term used here is not practical but practicable. If something is not practical it can be dismissed quite easily as impractical, but if something is not practicable it is not possible. For example, recruiting the whole cast of the Expendables film franchise as your test subjects may not be practical, but it is practicable. Similarly, flying to the moon for the weekend is not practical, but it may be practicable. In other words, dismissing something just because you don't feel like it is not an option here.

Finally, although this is no longer required in the standard, you will want to assess whether any criteria you had for the summative evaluation have been successfully met (see Section 2.3.7). Once all is said and done, and found to comply with requirements of IEC 62366-1, the usability of the medical device from a safety perspective can be presumed to be acceptable. Unless, of course, objective evidence (including post-production and post-market surveillance) to the contrary exists.

Clause 5.9 invokes the ISO 14971 standard on risk management via several non-binding notes. In such notes it is pointed out that also non-user-interface-related risk controls may be considered when assessing the necessity and practicality of further work to improve the safety of the interface, thereby branching over to ISO 14971 territory on risk control.

Further references to ISO 14971 are made to state that a) also residual risks stemming from the usability of the medical device are to be considered when assessing the final residual risks, and b) design changes resulting from usability engineering should lead to review of any new non-user-interface-related hazards and hazardous situations. Regarding summative evaluation of design changes, the non-binding annex A5.9 to the standard helpfully suggests that prior evaluation may be leaned on in terms of the parts that remain unchanged. This is an important practical concession and one we make use of in the roadmap we present in Section 10.

The non-binding annex to the standard, A5.7.3e, astutely points out that even when no actual use errors occur during testing, the test subjects may still experience discernible levels of difficulty in performing a task. This difficulty may be a sign of a weakness that will later result in a use error during real-world use. Even if the user does not comment on such difficulties, these may manifest themselves as, for example, pauses, hesitation, searching around the interface, or referring back to the accompanying documentation repeatedly. It may thus be in the manufacturer's best interest to investigate such weak signals and improve the interface.

For UOUP (see Section 2.5) Clause 5.9 is excluded in its entirety by annex C of IEC 62366-1. None of what was said above thus applies to UOUP components. Note that changes made to the UOUP strip the changed parts of any such exemption.

2.5 User interface of unknown provenance

In a political lasso-move worthy of Beth Dutton in the Yellowstone series, Clause 5.10 of the standard unceremoniously states that instead of Clauses 5.1 to 5.9 (see Sections 2.3.1 to 2.4.2), annex C of the standard is to be used in evaluating the usability of UOUP components. By the same swing of the rope, the annex in question is also deputized as a normative part of the standard (i.e., binding) so that its contents are to be interpreted in earnest and on par with the body of the standard. The annex does want to take one of the now dumb-struck clauses, Clause 5.1, with it but it is happy to dismiss the rest of the clauses.

We already defined UOUP in Section 1.18 as a medical-device user interface (or its part) which has been previously developed and may be lacking in its records for the usability engineering process as expected by the full IEC 62366-1 standard. Reading just the body of the standard it sounds like the term "previously" could equally well refer to something already developed by the manufacturer in the past or something now developed elsewhere by someone else using an unknown process. An interface developed before the publication of the IEC 62366-1 standard is naturally legacy, but according to annex C.1 so is an interface developed without adequate development records for IEC 62366-1. This would quite nicely align UOUP with the concept of software of unknown provenance (SOUP) as introduced by the IEC 62304 standard commonly used in software development for medical devices (see Juuso & Pöyhönen 2024).

However, from reading the now-normative annex C carefully it is clear that UOUP components are actually intended as a convenient, controllable mechanism for gradually bringing previously developed interfaces on par with present-day expectations and into the heard, so to speak. Once you perform changes on those legacy interfaces only the unchanged parts are said to remain classified as legacy – all the changed parts are now subject to the full IEC 62366-1 standard. This positions UOUP as something combining elements of both SOUP and legacy software as understood by IEC 62304 (see Juuso & Pöyhönen 2024). The term UOUP thus only sounds misleadingly similar to SOUP, but is not just "SOUP for UIs".

Crucially, UOUP isn't intended to be a permanent state of affairs for a component that is to be updated over time – each change should bring an UOUP component closer to full scrutiny by the standard, just like a piece of legacy software under IEC 62304 becomes less legacy with each update that takes place. With each change to an UOUP, or made because of the UOUP, you will have to apply the standard to a slightly bigger portion of your user interface. A SOUP, on the other hand, can remain forever and still get updates from its manufacturer. In other words, UOUP is a special pathway principally meant for addressing the usability of your own legacy interfaces, but it may also momentarily work for interface components sourced from outside your organization.

But what if you have a SOUP component that is used in the implementation of your user interface? The SOUP component gets updates and is handled as appropriate in the light of the IEC 62304 and ISO 13485 standards. Should we not care about the usability of that component? Or should we repeatedly apply the UOUP concept over the successive versions of the component?

In general, it will often not be possible to bring an externally developed SOUP component in-house. You will, therefore, have limited or no influence at all on how the UOUP-side of that SOUP might evolve to meet the expectations of the IEC 62366-1 standard for a good user interface. You could replace the UOUP with something of your own making, but this might be inferior to the UOUP in not only what it delivers to the users but also in terms of how safely and timely the module is then maintained. The FDA, for example, has occasionally shown concern over manufacturers trying to replicate services inhouse on a shoestring budget while some gigantic players are already maintaining those same services using vast resources. The same might be the case here if a manufacturer wrote their own Javascript library to handle graphical user interface elements already handled much better by some common and well-maintained library.

Conversely, you may be using a general multipurpose component that is used by tens of thousands of users in other domains, but you discover that you need a tweak in the component to use it in a medical setting. The manufacturer may, however, be reluctant to listen to your request as they are happy selling the component to the other tens of thousands of other users out there. This situation is akin to auditing your supplier and handing them a list of nonconformities to fix, only to hear a polite no as a response: you and your use are not significant enough in terms of dollars for them to act in any particular way. This does happen.

The answer to the above questions thus becomes muddied. The best solution to adopt will depend on the intricacies of each circumstance. One thing that is certain is that it is always the manufacturer who will have to weigh the facts of the matter, develop a sound rationale, and convince both themselves and the auditors on any choices they have made. Just because a vendor is deaf to your pleas doesn't mean that you too can grow a blind spot around the issue. It is up to you to find a way to keep the jigsaw together. When faced with such an impasse, our suggestion might be to adhere to SOUP processes and be influenced by UOUP as much as possible, but the soundness of this approach will depend on the particulars of the context.

Table 2.3 provides an overview of the key differences between UOUP, SOUP, and legacy software. For more discussion on the latter two please refer to Juuso & Pöyhönen 2024.

Annex C sweeps away IEC 62366-1 Clause 5 requirements in the case of UOUP with two exceptions: 1) a use case specification is always required as

Table 2.3 UOUP v. SOUP v. Legacy SW

	UOUP	*SOUP*	*Legacy SW*
Base standard	IEC 62366-1	IEC 62304	IEC 62304
Type of software	User interface	Anything except an entire software system	Any previously used medical-device software
Developer	Manufacturer themselves, or an external party	Usually an external party, possibly the manufacturer themselves	Usually the manufacturer themselves
Original development process according to the standard	Not expected	Not expected	Not expected
Later maintenance work according to the standard	Required for the changed parts	Not expected	Required for the changed parts
Changes subject to increased requirements from the standard	Yes	No	Yes

per Clause 5.1, and 2) anytime changes are made in UOUP components, or because of them, Clause 5 may come into play in its entirety regarding the changes. In place of other Clause 5 requirements, the annex sets a series of new binding requirements for UOUP components to meet. These are discussed in the following sections.

2.5.1 Use specification

IEC 62366-1 Clause 5.1 is the only subclause under Clause 5 to be retained for all use of UOUP, not just changes to such components or made because of them. The rest of the Clauses 5.2 through 5.9 are generally waived for UOUP, but importantly not for changes made on the UOUP or the device containing the UOUP.

No exceptions to Clause 5.1 are made by annex C2.1 of IEC 62366-1, and thus everything that was said above in Section 2.3.1 applies. The non-binding annex A of the standard offers further weight here by positioning the use specification as an essential document identifying key characteristics for the safe use of a medical device. According to the annex the use specification may in part be based on the accompanying documentation for the UOUP, and must in any case be consistent with that documentation.

2.5.2 Review of post-production information

Annex C2.2 insists the manufacturer review available post-production information (including complaints and field reports) for incidents and near-incidents, and by extension use errors. Any identified use errors that could lead to hazardous situations are to be entered into the usability engineering file. Similarly, when field information gives reason to believe that usability-related hazards or hazardous situations are implicated also these must be recorded and reviewed as instructed in the following (see Sections 2.5.3 and 2.5.4).

Note also that the non-binding annex A also chimes in here to suggest that more attention should be paid to the severity of the potential harms than the reported frequencies of the problems in the post-production data as that data may be subject to underreporting. There's a post that regularly makes its rounds on LinkedIn about a study of WWII fighter planes that returned to base riddled with bullet holes. Looking at the average placement of the bullet holes it could have been surmised that placing more armor in those areas might be a good idea – but equally logically the planes that did not return at all were probably hit in the remaining parts of the fuselage. To bring back more planes the added armor should probably be placed in those unrepresented areas. Counting losses in aircraft is no doubt easier than estimating unreported incidents, but the story serves as an anecdotal illustration of the point the standard makes here.

2.5.3 Review of usability-related hazards and hazardous situations

Annex C2.3 demands that the manufacturer review the risk analysis for the device containing UOUP, and ensures that any usability-related hazards and hazardous situations have been identified and documented. These can be documented in the usability engineering file, risk management file, or any other practical location, but a link from the usability engineering file is expected to be the appropriate location. Also, a record of this review is implicitly expected, if not explicitly required.

2.5.4 Review of risk control measures

Annex C2.4 demands that the manufacturer verify that adequate risk controls have been implemented for all the usability-related hazards and hazardous situations identified above (see Section 2.5.3). Furthermore, all risks must be reduced to an acceptable level as per the risk management policy. The verification of these requirements must be documented.

Crucially, note that if changes are required to reach an acceptable level of risks, these changes can't be considered UOUP and will be subject to the full weight of Clause 5 requirements.

2.5.5 Evaluation of residual risk

Based on all the information developed above, the manufacturer must re-evaluate the ISO 14971 overall residual risk status (Annex C2.5). This evaluation must be documented in the usability engineering file or the risk management file. While not stated by the standard, you can bet that you will also need to react to any discrepancy in the residual risk evaluations.

2.6 Non-safety qualities through usability

The lab-coat-wearing nerdy academics in us would have wanted to start this whole book with a section on non-safety-oriented goals around general usability and then move on to discuss what usability means in the more restricted context of medical devices. That would have made academic sense, but in the more practical everyday world we live in we felt that safety is paramount, and it should thus be discussed here first. Now that we have done that in the previous sections of this book, it's time to briefly look at how more general usability engineering may help us improve on safety without veering off in a completely different direction altogether. Think of Quentin Tarantino's "From Dusk Till Dawn" (1996), and the turn that adventure takes once entering the bar, or Angelina Jolie's magnificent "The Changeling" (2008) and the metamorphosis of the drama into a serial killer story. Here the story we hope to paint is one from just any flavor of usability to usability in the name of saving lives.

In medical devices, it is not enough to read "usability engineering" and think that all the issues can be solved with attention, time, photoshop, and some lipstick. However, seeing to the aesthetics in addition to safety may help to fashion a streamlined and more intuitive interaction narrative that in turn will help to foster better safety – or remove some distractions around safety at the very least. In designing a product, good general usability may help promote more efficient and more economical use of the device. In medical devices, the regulations and standards involved will shift the emphasis from streamlining use to fostering safety – from shaving off seconds from a lap time to preventing the racecar from crashing. In so doing we push general usability standards such as ISO 9241 aside in favor of standards such as ISO 14971, IEC 62366-1, and IEC 60601-1. Ultimately, both safety and efficiency may align as goals, and both can be used to build better medical devices.

2.6.1 Usability engineering in general

Usability outside the domain of medical devices has its own pantheon of heroes. For a comprehensive introduction into general usability engineering, you may wish to read Jakob Nielsen's landmark book "Usability Engineering" (1994, which is still the latest edition). The book is a hearty read, but still relevant three decades after its first printing. In his book, Nielsen provides a breathtaking ensemble of usability engineering projects and papers, and discusses his own theories into making sense out of what usability is and how good usability can be engineered. We won't steal his thunder here, but we do want to briefly introduce a few of the frameworks he discusses in detail in his book.

Firstly, Nielsen looks at what usability is, and what properties usability can be broken into. The five characteristics he presents are:

- **Learnability**
  The user interface should be as simple to learn as possible, preferably even intuitively usable without any learning required at all. This goal is shared by safety-oriented usability, where both documentation and training are understood to be necessary considerations but perhaps not actually always needed if the device is simple enough and safe enough to use without them.
- **Efficiency**
  The interface should be efficient to use so that productivity is increased. Note that no mention of safety is made here by Nielsen, but safety is very much at the heart of developing medical devices. Elsewhere in his book Nielsen does briefly discuss 'catastrophic errors' that result in, for example, lost or faulty work for the user, and how these might be addressed more rigorously than lesser errors. The microscope here is thus markedly different than in medical devices.

- **Memorability**
  Once learned, it should be hard to forget how to use the interface correctly. This characteristic is not commented on by the IEC 62366-1 standard on usability engineering for medical devices, but it is very much in the spirit of the standard. In practice training once given on the use of a medical device may need to be refreshed at some point and documentation may need to be improved based on post-market surveillance. Neither event should lessen the call to develop devices that are easy to use also when that use is intermittent, or the user is also burdened by the use of many other devices in the meantime.
- **Error rate**
  The interface should be such that errors in its use are rare and easy to recover from. This goal also holds for medical devices, but it may not always be a safe assumption that the user has the time, experience or fortitude to debug the devices – just consider a patient monitor that starts to complain about ethernet ports when the nurse attaches a new probe to the chest of the patient. Much will depend on the nature of the error and how clearly and practically it is presented to the user, of course.
- **Satisfaction**
  The interface should be pleasant and satisfying to use. Note that such goals are taken for granted in the context of medical devices and in any case relegated to playing second fiddle to safety.

All the above characteristics presented by Nielsen may be considered in the context of medical devices and they may add value to the development of medical devices. To focus on these, and just these, would however be a mistake. In the context of medical devices safety must take precedence and be closely followed by utility or the actual benefit being sought by the interface. If the above are the limit of our deliberations or if they somehow detract from our consideration of safety we will run the risk of not reaching conformity with our requirements and putting out a lesser medical device.

It is perhaps egotistical to suggest modification to something which has been proven to work for over 30 years, but when applied to medical-device development we would humbly suggest that the following characteristics also need to be part of our considerations.

- **Safety (including risks)**
  By now it is no secret that usability engineering is first and foremost about avoiding usability-related safety risks. We discussed this earlier in this section and will discuss it further for the remainder of this

book. Safety should thus be added to all aspirational lists presented by Nielsen.

- **Utility (including benefit, relationship to intended use)**
  Similar to how risks and benefits are discussed in the development of medical devices, beyond the topic of user interfaces, we should also look at what function the device performs for the user interacting with the interface. From the point of view of usability this should not just be about the benefit itself, but also how that function is made available to the user: the language and the concepts used, and the transparency of the controls and outcomes provided.

  All the above fits under the characteristics Nielsen presents, but discussing utility right out will emphasize how not everything is always "what you see is what you get" and not all handles on some advanced or abstract actions are equal. We feel that to squeeze benefit into the other characteristics obfuscates the link to risk management more than what is necessary. Also remember that any utility or benefit should be linked with the intended use of the device. We will talk more about benefit in Section 6.
- **Automation**
  Humans are good at innovative thought while machines are great at not losing their stamina while performing some repetitive tasks. When both humans and machines can perform some tasks, or when the exact point of the baton change between them can be tweaked, we should consider the impact our allocation of the task has on the usability of the overall device. Does one allocation make the process more transparent, or the concepts needed in the controls more, or less, intuitive to understand? In other words, when studying the usability of an interface packing more punch than just some simple mechanical lever clear to see – for example, one perhaps sending artificial intelligence on a quest to forage through our vast dataset – we should really be addressing usability considerations related to the automation we apply underneath the interface. See Sections 11.1 and 11.2 for more discussion of software and AI, respectively.
- **Preservation of the natural context**
  A simple reflex hammer used to check the reflexes of a patient sitting in front of the doctor wielding the hammer is straightforward to grasp: all is well if a hit results in a jerk of the knee quickly enough. With remote care there might be a longer lag involved between the fictional hit of the hammer and when a reaction is seen on the screen. With a more complex procedure the loss of true presence may have more profound

effects on the interpretation of the procedure. Not hearing the patient breathe in a labored way 500 miles away may cause a doctor to miss something in his chart. Not noticing something off in their recovery room, such as a background noise preventing the patient from sleeping, may also prevent optimal healing from taking place. In all these cases a part of the full context is partially lost, and this will have usability implications.

Not having the full context readily before you may change the interpretation of some piece of information, or some control provided by the interface. Considering whether a potential loss of the natural context is a relevant point to address for the usability of the device is now more poignant than when everything happened at no more than an arm's length. For more discussion of care processes and remote care see Section 11.5 and 11.6, respectively.

- **Changes occurring over time**
  Perhaps the most profound change affecting the usability of interfaces is that these may now change under the hood by design. No longer is it just wear and tear, or gradual deterioration, that may lead to unseen changes on a device while in use. If a scalpel loses its sharpness this is likely to be detected, but what if that scalpel received a software update between the time you last put it down and now pick it up again to apply to the skin of a new patient? Scalpels, for the most part, won't receive software updates but many other types of devices will. If the AI model inside a device, or just the plain old software code behind some interface, has changed this may or may not be immediately visible on the user interface itself.

  Both the advent of an update and its effects on how the user uses the device should not go by unaddressed. Usability engineering may have something to say on how the process of updates is arranged, how changes are marked up, and how those changes should be explained to the user in relatable, practical terms. The user too may mean both the individual interacting with the device and the responsible organization sanctioning its use (see Section 1.8). Addressing changes is a matter for both risk management and usability engineering in ensuring the user a) notices, b) understands the change having taken place, and c) if they are in the driver's seat for installing updates, both knows to do so and wants to do as part of their already busy work schedule. For more discussion of changes over time see Section 11.12.

The above additions are not meant to undermine the tried and tested model devised by Nielsen. Rather they are intended to highlight areas where devices may have evolved over the past few years, and where usability engineering might be brought to bear more than it has so far been. For more future trends perhaps affecting usability see Section 11.

Secondly, Nielsen proclaims ten tenets for achieving good usability which we feel succinctly dissect usability engineering also for the purposes of developing medical devices.

- **Use simple and natural dialogue**
  In other words, be both friendly and economical on what the device says to the user. The interface should not feel like it has ADHD and it should cut to the point instead of being unnecessarily verbose. The tenet applies to medical devices as well.
- **Speak the users' language**
  Building on the previous tenet, the terminology and content presented to the user should speak to them. Expecting all users to have engineering degrees or the drive to decipher user prompts like WWII codebreakers is not a smart stance to take. Instead, the language used should be familiar to the user. The tenet applies to medical devices as well.
- **Minimize the users' memory load**
  This could be characterized as an application of the 'need to know' maxim from spy movies: instead of relying on the user having the memory of an elephant or at least recalling everything six screens back, the interface should make information available where it is needed and when it is needed. Don't use the existence of a tutorial to gloss over a hard to use interface. The tenet applies to medical devices as well.
- **Be consistent**
  Just like when developing QMS procedures, don't feel like you must be a poet and come up with new terms or ways of saying the same thing. This will only confuse the reader. The same also applies to designing interactions and actions: these should always be presented in a similar way, involving the same expected steps (e.g., selecting a folder to save a file, then giving it a name, and finally clicking on 'save'). The tenet applies to medical devices as well.

- **Provide feedback**
  Most users are always wondering if the device is doing what they just told it to do, if it broke down, or when it will be ready with the task. The interface should thus provide meaningful status information to the user. Prompts, progress bars, and all sorts of indicators and flashing lights have been invented for just this reason. We might add that occasionally this feedback may also be quite unexpected to the developers, such as users getting accustomed to listening for the computer fan spinning up, and once gone on the next model – or the final production model succeeding an early demo device – missed. The tenet applies to medical devices as well.
- **Mark exits clearly**
  More than just closing a piece of software, users will also want to occasionally back out of a process. A good interface should make it clear how to abort or cancel actions when this is possible. The tenet applies to medical devices as well.
- **Support shortcuts**
  Nielsen believes experienced users will benefit from shortcuts and other power-user features that would only baffle first-time users. A good interface might make such shortcuts available to improve efficiency. The idea of shortcuts is naturally at odds with safety-obsessed medical device development as explained elsewhere in this section, but not necessarily incompatible if the manufacturer is willing to spend the extra effort in understanding how these shortcuts – including their accidental use – should be addressed along the full path of use.
- **Write good error messages**
  This tenet is linked with the item above calling for intuitive language to be used. The idea is, of course, that by reading the notification the user understands what has gone wrong. From the medical-device context this is a noble goal, and in some cases the user may adjust their approach based on what has gone wrong. However, in a real-world clinical setting the user may have quite limited time, desire, and familiarity with IT systems to react to the error in any way to debug the issue. For example, a message to check firewall settings may be much more appropriate for IT support staff than the nurse flicking between some views. The tenet applies to medical devices as well.
- **Prevent errors**
  Rather intuitively, Nielsen calls for all practical steps to be taken in preventing the occurrence of errors in the use of the interface. The tenet applies to medical devices as well.

- **Provide help and documentation**
  Ensure the help and documentation the user may want is available. The tenet applies to medical devices as well, although it may be too much to ask clinical users to always have such documentation at hand and enough time to dive in.

All the tenets introduced by Nielsen are compatible with the development of medical devices, but the essential topic of safety is ominously absent from the list. As long as the tenets are subjugated to safety, and not used to detract our attention from safety, all of the tenets are welcome in the development of medical devices and their interfaces. In the safety context we might inject the following additional tenets, based on the five new characteristics we campaigned for above.

- Consider safety first, last, and at every step in between
- Take a hard look at the benefit sought and its utility to the user
- Use automation where it improves safe usability
- Preserve the natural context as far as necessary
- Control changes transparently

Some of the lower-tier theories and graphs presented by Nielsen strike us a little bit "Pritchard-esque", as in the poem classification the Robin Williams character in "Dead Poets Society" (1989) has his students tear out in a move to appreciate the poems themselves. The comparison here is unfair as Nielsen's point is to assist in quantifying and classifying usability for the purpose of engineering better usability, not just to appreciate usability for the life-changing qualities it may bring. We thus recommend picking up a copy of Nielsen's book and making up your own mind on the utility of these models.

A wealth of other great books has also been written on usability engineering in general. A few of the most insightful reading experiences we would recommend include "GUI Bloopers" by Jeff Johnson (last updated in 2007), and "The Design of Everyday Things" by Don Norman (2013). You are, of course, spoilt for choice here, which is why we are profoundly grateful you chose to also read our book. If you have recommendations of your own, please don't hesitate to share those with us as we love to keep on learning. The book you hold in your hand is, however, the only one we have seen that anchors usability squarely on safety for its discussion.

2.6.2 Interpreting the standard beyond safety

In the previous section, we took a brief look at how popular tenets from general usability could be adapted for use in the development of safety-oriented usability for medical devices. Designing all medical devices in the same way you might a mobile game such as Pokémon Go is clearly not sufficient for fostering safety. But would it make sense to do the opposite: to expand the use of IEC 62366-1 usability engineering to non-safety goals?

The IEC 62366-1 standard on the application of usability to medical devices itself focuses squarely on the safety of those devices. A companion to the standard is published as a nonbinding technical report, IEC 62366-2, where the focus of usability engineering is cautiously broadened from the essential topic of safety to accuracy, completeness, efficiency, and user satisfaction. Of the four terms introduced efficiency is given the most attention by the standard. In fact, the term encompasses the first two characteristics of accuracy and completeness in its definition. Efficiency, it is suggested, is to be understood in relation to effectiveness (Clause 3.4) as the accuracy and completeness with which users achieve specified goals. This may also be linked with safety, such as in the case of a sterilization device that falls short in the completeness of the task entrusted to it, a pacemaker that has trouble keeping time, or an automatic defibrillator that lacks the requisite oomph. Note that keeping lap times with a stopwatch does not feature in any of these definitions. In medical devices even the concept of efficiency is mostly about thoroughness and reliability.

If we were to somehow forget about the safety aspects of usability engineering, the standard would look quite different. With safety out the window there would be little point to looking at safety-related risks, and thus the link to risk management would be gone (at least in the ISO 14971 sense). This would change the expectations of also summative usability evaluations where no tasks or scenarios would need to be studied. This would take a wrecking ball through most of the intricate relationships the standard has wanted to bring up between user interface design and risk management.

What would be left by this heavy-handed spring cleaning would be all the documentation of the user, use environment, and use of the device. Task analysis might still be quite practical and provide sufficient insight to be of use in the context of general usability. The mechanisms we might use to gather and assess user feedback might still look similar when our safety goggles were switched to some sleek sunglasses of user satisfaction. The usability engineering standard, however, does not carve out its place in the world by offering much advice on these tools, it only really calls for the consideration of safety and risks as part of any tools used here. The conclusion is that for all those occasions where accuracy, completeness, efficiency, and user satisfaction only affect the quality of the service somehow, but not safety in any way, the IEC 62366-1 standard will gladly sit on the sidelines.

2.6.3 Combining the two domains

In the previous section we carved away all safety-oriented sections of the usability engineering standard and ended up with little more than a set of documents describing the users and use of the device. All links to risk management were lost and our compass more or less lost the scent of the magnetic north.

Yet, it is possible to imagine a piece of medical-device software embedded somehow in a more general-purpose app. Think of a cardiac rhythm monitor embedded inside a generic wellness device. The calculation of some clinical parameters inside would need to be safe and effective in the sense of accuracy, but the step counter also residing in the same device might only get the ballpark correct. Both sets of features are used through the same display and the same buttons, so where does general usability switch over to safe usability? How should we draw the links across this divide to ensure nothing important is missed?

The low-hanging answer here is that how we describe the user, the environment, and the use of our device should not unnecessarily differ from one side of the fence to the other. The documentation suggested by IEC 62366-1 should most likely be adopted for use on both sides to ensure as little as possible is lost in any translations across the border.

The infinitely more difficult part of the question is how the links should be drawn, and how changes on one side should be seen on the other. Maintaining consistency and reflecting changes across the fence will get to be a chore. Perhaps the link should not be thought of as something automatic, but as a sequence of interleaving work where the effects of design changes are evaluated from both a safety and a sexiness mindset before each wave of changes is approved. Think of periodic safety update reports and scientific studies that might somehow trigger new risk management activities.

At the end of the day, clinicians, engineers, regulatory specialists, regulators, the management, sales, customers, users, and patients will have different goggles on when thinking about usability around medical devices. Even then, the usability standard itself asks for all these groups of people to somehow understand each other. The discussion will never be constant and forever flowing, but it may be punctuated by meaningful steps. As long as the sexy device is also safe to use and clinically effective – in other words the device is saliva-inducing in the way that loosens wallets instead of the way which threatens our health – all these groups of people will probably be satisfied. The good news here is that as long as you meet the safety requirements you are entirely free to attend to non-safety-oriented usability in any way you see fit. You will, however, need to ensure that those general usability activities don't adversely affect the safety of your device. That is not a requirement to be taken lightly.

Chapter 3

Usability Toolbox

Batman has his utility belt (see Figure 3.1*), Bond has an infinite supply of gadgets from Q, MacGyver was able to "macgyver" anything he needed from a piece of gum and a paperclip, and any lady's handbag famously holds everything but the kitchen sink. In usability engineering the toolbox is a little more predefined, but not without its fair share of adaptation to the needs of the particular device and investigation.*

In Section 2, we looked at the requirements placed on usability engineering for medical devices, their interfaces, and the interaction between these and the user. The discussion there was rooted in the normative language of the IEC 62366-1 standard and the requirements it made on our usability activities and documentation. The standard referred to a few tools it thought might be of interest to you as you dive into the real work expected of you, but it did not introduce these tools in any way. Instead, the separate guidance document on the standard, IEC 62366-2, was implicated as a practical extension of the standard and the source to consult for learning more on those tools. It may thus come as a bit of disappointment to read in the guidance that a person not yet familiar with such tools, task analysis in particular, should review the general literature on the topic. The guidance is a lengthy document of 100-plus pages offering welcome advice and experience-based wisdom, but frankly it is not comprehensive or detailed enough to really be the practical resource we want. We hope that this book is that.

In this section, we aim to dethrone the current guidance – but honor the hereditary accession rights of any future version of it – and discuss the practical layer we are all asked to create in accordance with the standard

DOI: 10.4324/9781003645702-3

Figure 3.1 UX-man's belt should have some document-sized pockets.

itself. As always, reading the guidance yourself will add some observations and no doubt provide further insight, but for once we are finding it hard not to call for the TR to be significantly revised. The TR should be half its current length and ten times as forthcoming and coherent in disseminating its wisdom. Be that as it may, this section will give you our rundown of the basic types of tools you might want to pack into your usability engineering toolbox.

3.1 Background user research

In films, method actors might conjure up an impenetrable bubble around their roles and refuse to come out of character for lunch breaks or even the duration of filming. Other actors may come up with intricate backstories to their characters to help them become these characters. Based on all the cast commentaries out there on DVD and Blu-ray, doing some form of background research on the topic of the film and the characters it involves is common practice. This background research may involve interviews with real people, watching archival footage of historical events or the people that shaped those events. In films the objective is to weave together a believable composite, a touching portrayal of people that may or may not have existed in real life or events that may or may not have taken place. In usability engineering these fictional characters might be called 'personas', which don't have to correspond to any known, interviewable user. More often than not, though, storytelling should be grounded closely with reality and what we know of the real users and use environments. In this context, personas should perhaps correspond to at least stereotypes of the real persons.

The reason we bring up movies here is not to promote the embracing of fiction, but to point out that the same kinds of research methods used by an actor in familiarizing themselves with their subject matter and the role they will be playing will, in fact, also be applicable to understanding the user. Research is research in acting, in science, and in any other field of industry serious about finding objective answers to its questions. Don't let the context of usability engineering and medical devices intimidate you.

Examples of the types of tools to use here in understanding the relevant field of medicine, healthcare workflows, and patient characteristics include:

- Literature surveys on the science and/or relevant field of medicine
- Surveys of relevant patient advocacy publications, magazines, newspapers, and social media
- Market surveys on both the type of device and its use environment
- Excursions to industry trade fairs and medical conferences
- Expert reviews and surveys
- Building experience with possible predicate devices, similar designs, or related healthcare workflows and the state of the art

The nonbinding guidance on the standard, IEC 62366-2, is quick to warn manufacturers against adopting market research as a surrogate to understanding the user via background user research (see Section 3.1). Rightly so, but market research may add valuable information to the puzzle we are attempting to understand.

In the case of an existing device now under investigation for possible improvement this step might also include:

- Analysis of customer and user feedback (including complaints)
- Analysis of post-market surveillance information (including any periodic safety update reports on the devices)
- Analysis of the change history of the device (including identification of similar past changes)
- Analysis of relevant nonconformities in products or processes (including identification of nonconformities identified via customer maintenance processes)
- Analysis of relevant product-specific risk management files (including identification of use-related risks, and any recent activity on these risks)
- Analysis of relevant process-specific risk management files (including any customer-facing processes from feedback to delivery, installation, and maintenance)

All the above information is based on activities required by ISO 13485. See our previous books for more discussion on these topics (Juuso 2022, Juuso & Pöyhönen 2024). The objective here is not to instruct additional hurdles to cross before usability engineering work could begin, but only to point out processes and information perhaps already being created in other quality management processes which may serve to inform usability engineering efforts.

The work done during background user research may feed into the development of the intended use (see Section 4.1.1), user profiles and user groups (see Section 4.1.2), and the description of the use environment (see Section 4.1.3).

3.2 Interviewing users

Talking with real live users should be a self-evident gold standard to understand their needs and circumstances. After all, users are the best authorities for telling you what they want from your device, what they would do with the device once they have it, and how they think it should be improved. This is even true when the users can't quite grasp the device you are pitching. Henry Ford is occasionally quoted as saying that if he had asked people what they wanted when he was developing the automobile, they would have said a faster horse. Lacking the concept of a car, the users could have perhaps spoken of horses, but the need to travel faster was either a genuine need or an invented one. It doesn't appear that Ford ever uttered those words, but the fact that almost everyone has heard the quote at one point or another possibly goes to show that it strikes a chord in telling us something about our need to relate innovations and present-day users.

Talking with folks is a good start, but it is pointless if you aren't able to deduce some lessons from all the interviews. Similarly, amassing vast amounts of research on users is not worth much if you can't then observe similarities and differences, slice the populations into user groups, and relate it all to the intended use you are working with. Occasionally insight may, of course, arise naturally out of the conversations, particularly if the volume of data you can acquire is low, but quality is high. At other times you might feel like the volume of data is fighting against you and masking the insight you hope is in there.

Some years ago, we were working with the Linguistic Atlas Project (LAP) in the US, at the time run by the University of Georgia. The LAP is a vast undertaking that spans decades of work by trained linguists who spent hours, days, and months interviewing people across the United States in an effort to map language across time and space. Using interview scripts, tape recorders, and hand-written notes the linguists documented thousands of interviews with awe-inspiring detail and ended up with one of the largest linguistic collections in existence in the world today. Combing through the data, formatting it, structuring it, analyzing it, and developing linguistic models based on it has been the main mission ever since – in addition to safe-keeping the unique data for future generations. My doctoral thesis, written at the intersection between linguistics and engineering, made use of this data, and went further than any previous research into enabling computer simulations based on the language, geography, and social characteristics of the speakers interviewed. To run the simulations based on the reel-to-reel audio tapes or the stacks of hand-written notes was beyond my skills and funding, but luckily several refined layers existed turning the data into keys, values, and even GPS coordinates more readily usable in computer simulations. The point of this long story was to say that instead of prose you actually want succinct, actionable observations and even key-value pairs of monitoring data from the use of your device. That way you might even see real reflections of the Appalachian Mountains in the data, based on how language changes around the mountain chain without any topographical information being encoded into the data. Just listening to the thousands of hours of interviews you probably wouldn't ever get there, or have the confidence to believe what you were hearing, but with appropriate coding and tools you can. The thesis is available online for free should you want to read more on what was done, how, and to what effect.

Recording prose is great for small-scale usability studies with fewer than maybe a dozen participants, but with much more participants than that you will start to increase the likelihood of masking out observations in repetition and noise. A good option to consider here is to use electronic evaluation forms that allow you to use multiple choice options and quickly pool the data you receive. The necessity of automating some aspects of observation and analysis, at least in part, will only grow as the size of your test group grows and as you later perform new evaluations and want to compare results. That said, don't drown out the unexpected by having too strict an approach and require using only predefined multiple-choice options at every turn.

Examples of the types of tools to use here include all the usual suspects of interrogating users:

- One-on-one interviews
- Group interviews
- Contextual inquiries (i.e., interviews conducted in the real setting)
- Questionnaires
- Digital observation (e.g., anonymized video tracking, eye tracking, mouse tracking, key logging)
- Expert panel groups
- Focus groups
- Hospital testbeds

One final, and perhaps somewhat obvious, thought on interviewing informants is that you should be weary of scripting responses too closely or feeling like you need to hurry the interview along. Some interviewees will veer off the topic freely every chance they get while others may seem resolute to not opening their mouths at all. Interviewing takes more skill than just mechanically crafting prompts to get replies, but the number one rule is always to listen to the person you are interviewing – be that in person or in writing. Listen for the answer you perhaps wanted, but not just that as you might be given nuggets of wisdom you didn't even know to dig for. The questions you ask should be relevant to the informant, and usually address their own spheres of everyday experience more than some hypothetical abstract realm of science fiction, but occasionally asking questions out of the ballpark will also yield unexpected gains. A study we once encountered – the exact citation of which unfortunately eludes us now – provided a halting statistic by claiming that around 70% of the information used by a doctor to form a diagnosis during a patient visit is actually reported by that patient themselves. The figure appears somewhat flawed in that little mention was made of the role of knowledge, information models, and probabilities of various possible ailments engraved into the doctor's brain via medical training and experience. Still, whether the statistic is strictly accurate or not, it anecdotally illustrates the importance of interviews in finding the right answers to the right questions quicker and in an all-round more humane way than just trial and error or by asking artificial intelligence (AI).

3.3 Observing users

Some exceptional films let us feel like we are flies on the wall witnessing history in the making or some weighty episode of personal drama unfold. Glengarry Glen Ross, based on a David Mamet play, is the prime example here. The film effortlessly transports us into the world of ordinary telephone salespeople, where the triumphs and tribulations of the winners and losers of that small microcosm pull our heartstrings with singular precision and immediate effect. The film does an excellent job of observing a group of people at their work, but it doesn't just document the tasks but also looks for the motivations, distractions, and interplay involved. The same should be our goal as we dive into observing users for usability engineering, be that with the intention of investigating the use of a specific device or discovering new opportunities for medical devices in existing workflows.

Examples of the types of tools to use here include all the usual suspects of stalking users:

- Contextual inquiries (i.e., interviews conducted in the real setting)
- Tag-along observation
- Hospital testbeds

3.4 Task analysis

There's an inspired skit in the Eddie Murphy film "Coming to America" where a restaurant patron calls the waiter over to taste the soup he's just been served. The waiter naturally inquires what's wrong with the soup? Is it too hot? Is it too cold? When the customer remains adamant that the waiter just taste the soup, the waiter relents only to discover that the spoon is missing. At this point the customer gloats gleefully "Ah-ha!". This unorthodox example serves to illustrate the point of task analysis: to discover all the steps involved in the overall task we want to accomplish so that we can ensure everything is in place and working as expected.

There is no single right way of doing task analysis correctly. A multitude of models, forms, and tools could be put forward as examples of how to divide a task into steps and how to record all the information involved therein in one form or another. So many in fact that the IEC 62366-2 guidance document begins its description of task analysis by suggesting the readers acquaint themselves with the literature out there first. This is a bit of a cold slap to the face of the reader, who like us, perhaps bought a copy of the document precisely to learn about the task analysis only mentioned in passing by the IEC 62366-1 standard itself. The guidance does have worthwhile lessons to report on task analysis, even if many of these are hidden between the lines.

The IEC 62366-1 standard itself does not elaborate on task analysis in any discernible detail. The standard only refers to task analysis in the context of Clause 5.2, a note related to that clause, and an entry in annex A related to it. The soft message here is that task analysis may serve as a possible way of analyzing usability and identifying use errors, and that for further discussion of the method we should refer to the guidance given in IEC 62366-2.

At the heart of it, task analysis is about observing and analyzing the steps involved in appropriate detail to allow us to truly capture and understand a task. The different templates out there may help you do this, or they may just distract you by splitting your attention and forcing you to record details that are unnecessary or unhelpful in your situation. Not to fear, we will go through ample practical advice in this section and then present our template for performing task analysis in Section 4.1.4.

3.4.1 Defining expected user tasks

The first step to task analysis is, of course, to identify the tasks we will be analyzing. These may be tasks we have noticed users performing with our device or some comparable device, or they may be the sort of tasks we envisage users tackling with our device once it exists as something beyond a drawing on a napkin.

The simplest way to get started with defining expected user tasks is to start with the defined functional requirements for the device. If no device yet exists, the first step to defining such requirements may be to conduct user observation of similar outcomes being sought in a setting where the future device is to be used (see Section 3.2) or take educated guesses to imagining the set of tasks if no comparable solutions truly exist. If a previous device does exist, this may be used to identify the tasks. Any cohesive high-level system requirements we might already have should act as an appropriate level of requirements to start with (see Juuso & Pöyhönen 2024).

The usability guidance, IEC 62366-2, speaks of anticipated user tasks only briefly in its Clause 8.2.2. Here the gentle message is that we can start anticipating user tasks even before any design for the device exists, and that these tasks may need to be refined throughout design and development activities. The guidance also mentions functional requirements as a potential stepping off point in addition to more creative exercises.

3.4.2 Refining user tasks

Once we have a rough idea of the overall bag of tasks we expect to be performed using our device, we can start going through those tasks to really flesh out the ins and outs of each task. We see task analysis as a multi-pass activity where you first define a task on a high level and then refine the description to break down each step as is practical. On the first pass you might record the grand strokes (steps 1,2,3, and so on) and then on the second pass add the sub steps (1.x, 2.x, 3.x, and so on). The guidance agrees and goes on to suggest that you might want to pay attention to the expectations placed on the user regarding perception, cognition, and action – something which is discussed in more detail in Section 3.5.

Once you feel confident you have the steps down on a level that is useful in your case, we would recommend taking a moment to look at a few further aspects of each step recording these as needed. The aspects we would consider here include:

- Users or user groups involved, so that we can ensure that if multiple user groups are in fact taking part in the task, we address each of them as is appropriate.
- Risks involved, so that we can begin to address the harms that may be associated with each step going wrong in some way.
- Benefits involved, so that if some specific benefits are associated with the step, we take note of this for later benefit-risk assessment.
- Use errors, so that we can address risks and risk controls appropriately for later development and risk management.

The above list is our own, the standard itself does not require anything specific to be recorded here, only that you understand the tasks expected of the user and are able to identify any associated use errors and harms. Section 4.1.4 provides our example of documenting tasks.

Something you may also consider here, but that is not required by the standard or suggested by the guidance, is capitalizing device functions in all-caps. This way you will start to see recurring actions in the text more easily. This suggestion is based on how film scripts are written drawing attention to key elements in the action by capitalizing them, and also on how we have grown accustomed to writing software test descriptions over the years, but it is by no means a requirement. We feel that using the same styling here will also make it that much easier to later develop tests against these task definitions. You could also use @-mentions to highlight users and user groups in your descriptions. Try it and use it if you like.

3.4.3 Identifying problems

The point of performing task analysis is not to get a nice stack of documents on how people should use an interface or a device. The point is to identify the prickly parts of the interface so that we can then apply the right amount of sandpaper to prevent any harm coming to the users. In other words, we captured a wealth of information on the users and the use via conducting analysis of the tasks users are expected to perform, but now it's time to take a look at that information with risk-management glasses on. The identification of use errors, harms, and risks is not technically a part of task analysis as understood by IEC 62366-2, but it is the very next step along the usability engineering path. In any case, task analysis is a practical tool for mapping use errors, risks, risk controls and use scenarios together in the context of all types of hazards (e.g., electrical shock, mechanical dangers, performance issues).

The expectations of the standard on identifying issues – be those in the form of hazards, hazardous situations, hazard-related use scenarios, use errors, or safety-related user interface characteristics – are discussed in Sections 2.3.2 – 2.3.4. We discuss the documentation and, more importantly, the information each of these lead to in Section 4.2.

3.5 PCA analysis

The guidance, IEC 62366-2, discusses a tool called PCA analysis to address the analysis of user perceptions, cognition, and actions. We object to calling this analysis 'PCA analysis' as we have come to regard that term as an acronym for principal component analysis used more widely in science. In the name of aligning our discussion here with the guidance we begrudgingly oblige, but really this should be called inherent impairment analysis, or something to that effect.

The objective of PCA analysis is to look at what specific demands we are placing on the user, in order for them to successfully perform the task asked of them. PCA gives us a way into asking ourselves whether there might be some foreseeable issues in the user being able to see, hear, comprehend, and take action as we prescribe. Think of the scrawny little runt of a man asked to become Captain America in that Marvel movie, or any of the unlikely heroes in films asked to grow into the bigger boots put in front of them by fate, but maybe don't require similar growth journeys from your users. Some of them might grow into those boots, but with medical devices the users should not be expected to have superhuman powers even if they are heroes to their patients.

In Section 4.1.4 we present our take on a form of task analysis that also begins to address impairments as part of analyzing the steps. You may decide to adopt our approach, perform separate PCA analysis, or decide that your users are sufficiently homogenous and your tasks simple enough that a PCA analysis does not offer valuable insight. You are not required to conduct PCA analysis, but you should if it can help you make your devices safer.

3.6 Function analysis

IEC 62366-2 briefly talks of a tool called 'function analysis', which in essence addresses the division of tasks between human users and some form of automation in the device. The idea is that humans have strengths and weaknesses, and so do machines, be those mechanical or computational in nature. The objective of function analysis is therefore to assist the designers in making the right allocation for each function. We touched on a similar notion in Section 2.6 when discussing Nielsens theories.

In films the infliction point between man and machine is taken to spectacular extremes in sci-fi stories such as Terminator 2, 2001, and Bladerunner. Machines may have difficulties learning the human trait of smiling, but they may decide all by themselves that a little apocalypse is in order to perform the task we gave them – or they may drop the task altogether and learn to love other electronic sheep instead. There are plenty of examples of men behaving like machines, too, from The Jackal to The Accountant, to name but a few of the most mercenary films out there without verging on the serial killer films.

The goal of function analysis is, of course, more down to earth than any of these examples. Function analysis only attempts to ensure that we play to the strengths of both man and machine in defining the functionality of our device and its interface. The likes of ChatGPT may appear to have led us to murkier waters in this regard, but still the axis from creativity and adaptability to speed and repetition ought to serve us well in figuring out who should take care of which subtask.

Interestingly the IEC 62366-2 guidance document also touches on the fit between man and machine by pointing out a few problem cases. Here the guidance suggests that human users may lose their situational awareness by interacting more with the device than their surroundings, or get played out of the loop by an overly fast device or by failing to notice some critical small change in the data. The document does not go into any detail on resolving or preventing such issues, it only suggests that the assignment of functions between man and machine might be at fault. The issue may, however, have more to do with usability engineering around the features at this crossover point than the mere assignment of functions. The difference here is one of pronouncing potatoes, of course. What matters is that the functions are assigned appropriately, and the changeover is designed safely. If you think of the use of your device as a relay race with robots, when and where should you change the baton?

3.7 Prototyping

The final summative usability evaluation should be done using the final product and representative users performing the tasks implicated in the intended use of the device. If the evaluation falls short of these expectations, you will have a harder time making the case that the evaluation has a place in the overall validation of your device as per ISO 13485. This, however, doesn't mean that you couldn't start early and couldn't start roughly with prototyping in your formative usability evaluations.

A few decades ago, we intuitively created Photoshop mock-ups of the graphical user interfaces we wanted to create. This was our way of thinking about the task at hand and using a tool we were familiar with to tease out a thought-out iteration of what the end product could look like. We liked working in Photoshop, copy-pasting elements as needed and ending up with concept images that people thought were screenshots of finished software. In retrospect the approach had flaws.

Firstly, it would have been faster to go for a wireframe instead of a pixel-perfect concept image. Relying on copy-pasted elements made earlier, either during the same project or a previous project, may have limited our thinking at times as it was easier to reuse something than to think further outside the box. It may also have made others more reluctant to offer comments or alternative ideas as the concept images looked like finished products. Secondly, had we not also been the ones implementing the approved concepts, it could have also happened that something put into the drawings was followed more closely than intended. Jeff Johnson, the author of the book "GUI Bloopers" (Morgan Kauffman 2000), for example, tells a cautionary tale where he used characters from the Flintstones as profiles in his GUI design and later was surprised to see these implemented as hard presets in the final software. Later we have also come to realize that using a wireframe instead of a more intricate bitmap image may be more useful during software testing where we want to ensure that the design was implemented as intended, but maybe don't want every pixel to be the same.

A benefit we enjoyed from using photoshopped images was that we could create simple click-through animations of how the software could be used, and show these to our supervisors. Toggling the visibility of the layers and layer groups in Photoshop we could create successive images where menu options and interface elements changed as needed to show a narrative through the interface. This was quite cool in the early 2000s but is now better done using other tools. Nowadays there are better, dedicated tools for experimenting with user interface concepts, but the same incremental approach to thinking out and iterating an interface via mock-ups holds. It may also be the case that an early paper prototype or other quick-and-dirty mock-up has less noise around the very ideas you are trying to put forward, and thus is the best prop to use in evaluating your overall idea. Later prototypes can be more refined and offer more insight on the use of the device, but you don't have to wait until the end of your development project to get started with usability engineering. In fact, you shouldn't.

Examples of the types of tools to use here include:

- Paper prototypes using wireframes, concept images, and even post-it notes
- Simulations and animations
- Interactive mock-ups
- Alpha- and Beta-versions of software

3.8 Testing similar devices

In one of the very first meetings I ever attended in the bombshell bunker of a huge telecommunications manufacturer I was surprised to see all the bigwigs flash out devices from their competitors. One executive saw my surprise and commented in a friendly, matter-of-fact manner that it is only natural to test out the competition. The same goes for the medical-device industry.

Although it seems unlikely that a manufacturer would want to perform costly usability engineering activities on a competing device, this might lead to deep insight on how to differentiate from the competitors and also what marks the device has to hit in order to compete in the first place. A full-blown usability engineering effort may not be required, but some consideration ought to be given to other similar devices out there, particularly if they happen to tick the predicate-device box for regulatory submissions. All this information may be valuable input into your clinical evidence activities as well as regulatory activities.

Examples of the types of tools to use here include:

- Ad-hoc test drive of a similar device
- Usability tests with similar devices
- A-and-B usability tests involving both a device and similar devices

3.9 Training

Wait, what? Training? Isn't teaching for those who can't "do"? Or training meant for patching up issues when you couldn't fix them in the first place? Some form of training material or sessions are perhaps called for, but don't plan on doing special training before usability evaluations and then neglect to do the same with all users before they use your device. That might be called doping, and you want to keep that at least an arm's length away from the tour bus we are on. It may be one small bag of blood plasma for a man, but still a pretty big deal for mankind when it comes out – to mix up one Armstrong with another here.

IEC 62366-2 Clause 15.5 addresses the development of training and training materials. The jumping off point for thinking about training is to understand the device, the user groups using it, and the extent of the training required. In its simplest form, training can be just a quick start guide to a simple, low-risk device. At the opposite end of the scale training can amount to some form of a periodically refreshed certification needed to use the device. Whatever level of training is thought appropriate for the device, the training should also be observed when conducting usability evaluations. This is particularly important if the training material forms a part of the overall risk control activities for device, is considered information for safety, and the effectiveness of which may thus require verification.

The material should also consider its audience and, as all accompanying information for a medical device, be appropriate to that audience in both content and language. In the same vein, video tutorials or 3D models of some features of anatomy or physical device may also be appropriate for training purposes. Such complex artifacts are ill-suited for quick reference use, but for leisurely training they may be very useful. I still remember being amazed during my first day of military training years ago when the instructor pulled up a YouTube video while explaining the function of an assault rifle. The video was both illustrative and memorable for sure, but as it appeared to have been shot by a weapons enthusiast it likely wasn't meant for military training of any sorts – and you don't want to resort to watching the video for instruction in the heat of battle. Today, such use of video is commonplace, but in medical devices the manufacturer should, of course, assert a level of control over the material being used and ensure its availability over the long term.

We would also add that the training could use both educational content and some material for easy verification of those educational goals having been met. The verification may come in the form of self-assessment questions with provided model answers or involve a separate check of the answers by, for example, the responsible organization (see Section 1.7). The material here could consist of printable PDFs or a link to the manufacturer's educational portal.

Note also that ISO 13485 expects the manufacturer to define any required training and ensure its later availability at the time of releasing a product. In other words, the usability engineering guidance suggests this should be the case, but ISO 13485 full-on requires this to be the case. As for how this training is conducted is left to the manufacturer to decide, plan, and execute – possibly with the aid of the responsible organization. The guidance, IEC 62366-2, is frustratingly vague and repetitive in its discussion of training, but as neither it or the standard go into great detail on the matter, you are free to address your training needs according to what makes the most sense for your device and its use. In the end, to the maximum degree possible, you should endeavor to make the device safe to use without any training, and then use training to help your users get on the right footing in using it. No training – barring a Manchurian candidate -level of mind control employed by the training – will be remembered forever, but if your device can be used both in a smart, efficient way and a dumb, haphazard way the users may be more inclined to the former after being shown the light. We, however, can't assume this to always be the case.

3.10 Accompanying documentation

In addition to training, the information material that comes with your device deserves its own mention. This may consist of the user manual, installation instructions, a quick start guide, or any documentation you deem necessary. ISO 13485 expects to know if such materials are needed to use your product and for you to ensure that they are available if needed.

The mean way of characterizing accompanying documentation is that they are the dumping ground of any additional instructions you feel you need to rid yourself of blame when something goes wrong with the use of your device, or that they are a form of paperweight literature to be deposited in the janitors closet at the hospital. They should, of course, be neither.

Accompanying documentation describes and instructs the use of your device in its intended use environment, in language readily understood by its intended users. In fact, such documentation is viewed as part of the user interface of your device by, for example, IEC 62366-2. As such, it almost inevitably becomes a part of the usability engineering activities around your device. Residual risks should be described in this documentation along with any other information for safety – and the effectiveness of this all should be a part of your summative assessment.

In other words, accompanying information should be designed just as carefully as your overall medical device. It should not be the accumulated leftovers from your clipboard in Word after all other information has found a place somewhere in your graphical user interface. It should not be included in the box with your device just to make the buyer feel a bit better about the substantiality of their purchase.

We will discuss accompanying documentation in more detail in Section 4.4.

3.11 Desktop evaluation

The nonbinding guidance to the standard, IEC 62366-2, introduces the concept of desktop evaluation as an alternative to usability studies. This is to be understood as the drydock version of the sea trials of a ship and suggested only for minor modifications of an already otherwise evaluated interface. The expectation is that such modifications don't increase use-related risks and can't be expected to create new use difficulties.

The term implies that the usability evaluation takes place without test subjects and without usability tests, as a document-based exercise. A solid rationale for relying on desktop evaluations instead of summative evaluation should always be recorded.

3.12 Formative evaluation

As discussed above in Section 2.4.1, formative evaluations are a discretionary tool in the manufacturer's toolbox. These can be run as needed to form an assessment of the usability of some aspect of the user interface, or even as a dry run for greenlighting the ultimate summative assessment of the whole.

Formative evaluations can take the form of, for example, reviews, surveys, interviews, or shallow-end usability tests. IEC 62366-2 Clause 16.2.2 also presents the practice of periodically reviewing the interfaces under development against the requirements set for them, as well as against any applicable usability principles, as a form of formative evaluation. Similarly, although not suggested here by the guidance, even some form of peer reviews conducted on the interface code might qualify as part of formative evaluation.

Going beyond just evolving code, the evaluation might involve some wireframe, concept image, or simulation of the user interface. Ultimately the evaluations might involve a functioning prototype of the interface or the finished, but as-of-yet unreleased product. In filmmaking, a formative evaluation might take the form of storyboards or animatics being shown to the crew or some other early test group. The summative evaluation, then, would be a full-blown test screening of the finished film before the fate of a theatrical release or a marketing budget is decided on.

In the domain of medical devices, formative evaluations may look at the whole interface or just a part of it, and the version of the interface may just be code, designs, or the finished thing itself. Given this ambiguity, or non-specialization, it is no wonder that formative evaluations are suggested to take place all throughout the design and development of the device. Formative evaluations really are the all-terrain approach to usability engineering.

3.13 Summative evaluation

Summative evaluations are the big-ticket item in usability engineering, and the one thing everyone knows to ask for when discussing your usability activities. These are the shark you'd expect to see when walking into a screening of Jaws or the Jedi you want to see act his revenge in a film called The Return of the Jedi. Bruce Willis may have failed to die in Die Hard, but with that exception, the advertised main attraction of a film is expected to materialize sooner or later. This is the same with summative evaluations, which are the leading stars in any usability engineering effort.

Interestingly, the nonbinding guidance on the standard, IEC 62366-2, takes a somewhat Highlander-like approach to summative evaluations in its clause 17.2: in the end there can be only one summative evaluation. If you began a summative evaluation and it failed to give your interface a clean bill of health, for whatever reason, it is suggested to be stripped of its title and demoted to a formative evaluation leading up to a new summative assessment. This is a rather harsh code to live by and we would instead suggest analyzing the results of the evaluation and acting accordingly, not just redoing an evaluation by reflex (see discussion of reacting to the results below). Otherwise, you might just be rolling the dice again, and perhaps even loading the dice in the process by relaxing the rules somehow. Afterall, the guidance only attempts to be helpful while the requirements are set in the standard itself. Performing a summative evaluation in parts is also allowed by the standard.

In a baffling remark to Clause 17.4, the guidance goes even further to suggest that if a follow-up summative evaluation is conducted after a successful summative evaluation that first evaluation is redefined as a formative evaluation. This to us is an exercise in extreme bureaucracy and has the potential of inciting uncertainty and disrupting traceability. This rebranding of already approved and probably externally referenced documentation is a plain bad idea. After all, where would be if we stripped Arnold of the title of Terminator after the first film and referred to him as, for example, the Prototype or just 'the machine without a name' as he battled the new liquid-based T-1000 foe in the second film. The standard itself does not make any such demands of automatically scrapping previous summative evaluations, and it is only the separate guidance that appears to be smitten with boyish fatalism here.

We already looked at the actual requirements of summative evaluations in Section 2.4.2, so here we will concentrate on how to conduct such evaluations. The nonbinding guidance, IEC 62366-2, discusses summative evaluations in detail in its Clause 17.

Running a summative usability test

Your summative evaluation may address all the tasks associated with your hazard-related use scenarios, or just some subset of these based on a valid, documented rationale (see Section 2.3.5). Note, too, that only addressing tasks associated with hazards and harms is expected, not all the user tasks you may have identified during development.

According to IEC 62366-2, summative evaluations usually involve conducting usability tests in simulated conditions. Creating a realistic simulated use environment without affecting the real clinical environment in any way may be a trick to achieve. Hospital test beds and other simulated development and training environments run by various healthcare facilities may be a great solution to use here. The guidance also notes that occasionally conducting true usability tests may be difficult or unethical. In such circumstances other methods (e.g., expert reviews) may be used, but an acceptable rationale should be recorded in the planning of the evaluations (see Section 2.3.7). The examples of usability tests given by the guidance allow using a mannequin instead of a patient, and taking the needle or the drug out of an injection device. All these examples are somewhat eyepopping after reading in the standard that a representative product must be used, and that the users involved in the study should be representative of the intended users. It all depends on your aims for the evaluation, of course. Using a manikin as your user and sticking post-it notes on it to mark injection sites begins to sound quite artificial, but so does injecting an inanimate patient manikin repeatedly with some precious real biopharmaceutical product. Your evaluation plan should make it clear what you're trying to achieve and how, and your activities should then match your plan as closely as possible. Any deviations from the plan should naturally be recorded and presented in the report of the evaluation.

The IEC 62366-2 guidance takes a stab at considering the effects of the size of the test group. In its Annex K it offers a mechanism for calculating the probability of a certain use error being encountered given a test group size (N) and a guesstimated likelihood of an individual test user encountering that particular issue. The nonbinding guidance ends up implying that N=15 is a quite reasonable target size for the test group, as was also suggested above in Section 1.3. The implication here is that use errors with a probability of less than 4,5% have a 50:50 chance of being unearthed in tests with a group of 15 subjects. For a test group of 5 subjects this blackout zone increases to all use errors with a probability of less than 13%, and for a test group of just 1 subject the same chance of slipping by undetected would apply to all use errors with a probability of occurrence under 50%. At the very least, Annex K offers good food for thought.

Collecting data

The expectation is that you'll have something more to show from summative evaluations besides just a final pass/fail decision. The standard doesn't go into detail on what data it expects to see out of the evaluations, only that you must identify use errors and close calls from the records.

The detailed planning for the evaluations should already specify very nicely what the test users were expected to do, where, how, and to what effect. The observations you make during the tests as well as the subjective remarks offered by the test subjects themselves will now add to the body of data generated by the evaluation.

Summative evaluations are qualitative in nature almost by necessity as a quantitative evaluation is beyond the scale of a traditional usability study. For addressing usability during post-market surveillance activities, it may, however, be worthwhile considering if real-world data might somehow offer sufficient large-scale data for a quantitative investigation. The standard or the associated guidance don't discuss the use of real-world data, but as that flavor of data is becoming more accepted even in building the clinical evidence for medical devices, we should perhaps expect it to increasingly feature in usability assessments too. For now, only simple statistics (e.g., user error counts, close call counts, task execution lengths) are expected.

In addition to the notes a study moderator might record in their notepad, you may also consider using automatic tracking software (e.g., activity logs, eye motion tracking, mouse tracking) and video recording from the test situation. You may also be surprised by how much automated processing can be done on such data recorded – even as far as detecting heart rates from a facial video to name but one very real modern-day sci-fi application. The nonbinding guidance considers observations as objective and comments volunteered by subjects as subjective, which is a somewhat simplistic view to take. In practice all observations are subjective, and user comments may actually be the most objective and telling observations about a device, and only automated measurements are likely to be reasonably objective by nature, but none of this detracts from the value of observations in usability evaluations.

In addition to creating records during the tests themselves, the guidance suggests the use of post-testing interviews. These will give the test subjects an opportunity to comment on the tasks and offer valuable further information to investigate use errors, close calls, and use difficulties. The use of post-testing interviews is suggested as a superior alternative to asking the test subjects to think out aloud while performing the tasks as that might increase the artificial nature of the tests. Video recordings may be used to a good effect during such interviews, but unfortunately video recording may itself make the tests feel more awkward and artificial for the subjects.

Analyzing data

As rigorous, objective, penetrating, and truthful as we want to see all usability evaluations be, they will always be subjective to some extent. They will always be a sampling of the use of the device. People wrote the tests, executed the tests, recorded the observations, and aspired to make impartial assessments of it all. The way the tests were written, understood, executed, observed, probed, commented on, and analyzed is unquestionably the result of work by humans with particular traits, inclinations, skillsets, and educational backgrounds. Having this natural neural network to rely on is the key to making sense out of the signals revealed by the usability evaluations. That said, we should attempt to make that neural network as wide, varied, and comprehensive as is appropriate.

Everyone knows that the generally desired outcome of the test is a greenlight for the product to proceed along its path to release, but this should not be at the cost of passing an unsafe device. Fixing any issues now will be both safer and cheaper than later on when the product is launched and real users are exposed to it. The outcome of the summative evaluation should thus not be preordained or the work so pressurized that use errors and potential issues go unhandled.

The nonbinding guidance to the standard rightly points out that occasionally a use error may be the result of the test subject willfully ignoring instructions. In such cases the device may not be at fault and no adjustment of the user interface may be called for. It should, however, not be your go-to position to blame the user, nor should it be a lightly invoked trump card. Each encountered issue should be carefully investigated and the implicated links to design and development, as well as risk management followed to arrive at the true root causes behind the issues and make the necessary adjustments. The summative evaluation may also cause updates to the risk management of the device.

The final expectation is that after completing the summative evaluation you as the manufacturer are able to say that no further improvement of the user interface is necessary or practicable (IEC 62366-2 Clause 17.1). The outcome of the final summative evaluation should thus show that the interface is not susceptible to potentially harmful use errors. The evaluations may, however, show that use errors did occur or that a series of close calls were observed. In this case risk management methods may be used to determine whether further improvement of the user interface is necessary and practicable (i.e., possible), or if the residual risks should be accepted as is in view of the benefit associated with the device. In other words, the right course of action will depend on the exact nature of the encountered issues, and not be a knee jerk action written into the standard.

If the evaluations dredged up the super-villain Penguin we should probably light up the Bat-signal, but in case of some lesser sidekick minion we might act by invoking possibly constraining the intended use or revising accompanying information. Use this realization to refrain from brushing issues under the carpet or scripting the final usability evaluation beforehand: issues will appear, but they may not be showstoppers if you assess them appropriately and act accordingly.

Reacting on the results

The results of the summative evaluation should be considered in the overall risk management of the medical device, synchronized both ways, and perhaps feature in the overall benefit-risk assessment of the device.

In principle, the choice of steps after a summative evaluation are:

- **The user interface was cleared as is**
  No further work on the user interface is deemed necessary or practicable. A residual risk evaluation as per ISO 14971 is expected next.
- **The evaluation failed to clear the user interface**
  Further improvement of the user interface is required, and this will be followed by a new summative evaluation. Joint work between usability engineering and risk management may be appropriate already at this stage.

If new use errors, hazards, hazardous situations or hazard-related use scenarios were detected during the summative evaluation we need to take a long careful look at them and redo work as appropriate. Similarly, if we are forced to later revisit a previously cleared summative evaluation as a result of detected nonconformities, we may need to do much more than just update the design and risk management. This is, however, addressed via nonconformity handling and is the topic of our previous two books (Juuso 2022, Juuso & Pöyhönen 2024). Usability engineering will then feature in those investigations as is relevant.

3.14 Problem resolution

The standard itself, IEC 62366-1, is all about identifying issues and use errors in the user interface, and ironing these out before the product ships. The standard does not have much to say on reacting to issues detected later when the product is in real use by real users. In fact, it doesn't even directly address the investigation of complaints by users or the appropriate handling of incremental changes over time. Instead, it appears to take it for granted that, as problems pop up, we rerun usability engineering efforts in their entirety, or at least on some appropriate level, and ultimately obtain a new summative evaluation. This is a nice theory, but in practice perhaps quite unlikely to happen with every small bug fix, maintenance update, and new version of a product.

The guidance on the standard, IEC 62366-2, jumps in to fill the holes here and discusses reacting to problem reports from users, soliciting more information to investigate issues via specific sample questions, performing root-cause analysis, and addressing the issues appropriately. In such discussion the guidance lands firmly in the domain of ISO 13485 customer feedback handling (see Juuso 2022), but it somewhat confusingly calls this 'post-production review and analysis' here so as the emphasize a delayed, retrospective approach. The gist is that use errors are likely to be underreported as users are busy with other things, and we should put real effort into understanding the reports we do get as manufacturers.

Neither the standard nor the guidance calls for a separate usability-flavored process of complaint handling or problem resolution. Instead, we should look to involve a usability engineering angle in the appropriate customer feedback handling, product maintenance, and risk management processes (see Section 9). The term used here in the heading, 'problem resolution', is borrowed from the software development standard IEC 62304, but we feel that it is appropriate to the discussion here and part of the reactive aspect of usability engineering that is missing from the usability standard. We will return to this topic later in the book.

Chapter 4

Medical-Device Documentation

The user interface of your medical device is not just the software-based graphical user interface (GUI) it might come with, but instead it is all the levers, pulleys, hoses, cables, buttons, scrollbars and other constructs left open to user interpretation and manipulation. Accordingly, the usability engineering file is the documentation developed to specify, design, implement, test, and evaluate that interface. The documentation here will not be a separate silo of archival records locked away in a vault, as perhaps illustrated by Figure 4.1, *but will instead have significant overlap with your overall living body of medical-device documents. For this reason, we will discuss the whole body of such documentation here.*

We discussed medical-device documentation in detail in our previous book on medical-grade software development (Juuso & Pöyhönen 2024) where we looked at all the major standards governing medical software documentation. In the book, we discussed usability engineering, but only from a high-up perch focusing on formative and summative evaluation reports. In this section we'll look at the detailed expectations that usability engineering places on our software documentation and look at both the usability engineering file and the user-facing documentation accompanying a released piece of software.

The discussion in this section is structured around the following subtopics:

DOI: 10.4324/9781003645702-4

Figure 4.1 What happens to the usability of the usability engineering file if it's entombed in a vault?

- The usability engineering file
- Risk management documentation
- Other development documentation
- Accompanying documentation
- Post-production documentation

In the following, we'll discuss each type of documentation and its relation to usability engineering. For more discussion of how documented information, documents, and records differ from each other and the subtle difference in how these are understood by ISO 9000 and ISO 13485 please see our first book (Juuso 2022). For our discussion here in this section such distinctions are purely academic.

4.1 The usability engineering file

In the discussion above, we have already made numerous references to a by-now mythical usability engineering file. The IEC 62366-1 standard, too, makes several references to this body of documents when it wants to see this or that record retained for later use or for posterity in general. The related nonbinding guidance, IEC 62366-2, also speaks of the usability engineering file at length.

IEC 62366-1 Clause 3.18 defines the usability engineering file as a set of documentation (including records and documents) that are produced by following the usability engineering process instructed in the standard. Clause 4.2 of the standard takes a quick second stab at defining the concept but amounts to little more than a statement of the fact that the file consists of records and documents created by the usability engineering process, and that these may in fact form parts of other files such as those included in your medical device file. The file may thus be one neat dossier, or it may be a network of all the necessary information contained in your medical device documentation. The nonbinding annex A3.18 agrees with this interpretation although it really doesn't get to have a strong voice on the matter.

Neither clause 3.18 nor 4.2 attempt to identify an exhaustive list of typical documentation to include in the file, but from reading the body of the standard these can be inferred to often include the following:

- Product realization documentation
- Risk management documentation
- Instructions for use
- Instructions for delivery, storage, installation, maintenance, repair, and decommission

Diving one level deeper, the IEC 62366-1 standard implies that it expects to see the following discussed in the usability engineering file:

- Use specification (Clauses 5.1, 5.3)
- User profiles (Clauses 5.1, 5.3, 5.7.1)
- Safety-related user interface characteristics (Clause 5.2)
- User errors (Clauses 5.2, 5.3)
- Known or foreseeable hazards and hazardous situations (Clause 5.3), including those learned regarding similar existing medical-device user interfaces (Clause 5.3)
- Abnormal use conditions (Clause 1, non-binding note to Clause 5.3)
- Hazard-related use scenarios (Clause 5.4)
- Summary of the selection scheme for use scenarios to include in the summative evaluation, including rationale for its use, and the results of its application (Clause 5.5)

- User interface specification (Clause 5.6)
- User interface evaluation plan (Clause 5.7)
- Formative evaluation reports (Clause 5.7)
- Summative evaluation report (Clause 5.7)
- Training and training materials (Clause 5.8)

Table 4.1 sorts the documentation into groups by type. The table also provides references to the sections of this book where each document is discussed in more detail. Note that easements on the documentation requirements may be made for UOUP components, as discussed above in Section 2.5.

Table 4.1 Usability documentation by type

Category	*Required documentation*	*Optional documentation*
Design documentation	■ Use specification (4.1.5) ■ User profiles (4.1.2) ■ User errors (4.2.3) ■ Safety-related user interface characteristics (4.2.4) ■ User interface specification (4.1.6)	■ Literature surveys, other background research (4.3) ■ Use environments (4.1.3) ■ Task analysis (4.1.4)
Testing documentation	■ User interface evaluation plan (4.1.7) ■ Formative and summative evaluation reports (4.1.8) ■ Test cases and procedures for each user group including their acceptance criteria	■ Usability evaluation report (4.1.8) ■ Photos, videos, and signal captures of usability tests (3.13) ■ Test results and findings for further usability analysis, as relevant
Risk documentation	■ Known or foreseeable hazards and hazardous situations (4.2.1) ■ Hazard-related use scenarios (4.2.2)	■ Literature surveys, other background research (4.3) ■ Abnormal use conditions
Accompanying documentation	■ Instructions for use (4.4) ■ Instructions for delivery, storage, installation, maintenance, repair, and decommission (4.4) ■ Release notes (4.4) ■ Residual anomalies (4.4) ■ Training and training materials (4.4)	■ Instructional videos ■ Tutorials ■ Quick guides ■ Self-assessment tests

Unless you like churning your milkshakes into butter, don't reinvent the wheel here or clone documentation created elsewhere in your quality management processes. Instead, develop a smart pipeline for product realization and ensure that this feeds in as appropriate to your usability engineering activities, and vice versa. We will discuss integration with your overall QMS in Section 9.

4.1.1 Intended use

Developing the intended use (or intended purpose as it's known in Europe) is not a topic addressed by the standard or its guidance, IEC 62366-1 or IEC 62366-2, respectively. There is work taking place at ISO on such guidance as we speak, but for the most part, this is a topic best left to the applicable regulations to govern. Yet, this is a fundamental piece of the overall puzzle to developing medical devices. In many ways, the intended use is the anchor of the medical devices to regulations and standards. All medical-device qualification ("is my device a medical device?") and classification ("what class of medical device is my device?") are based on the intended purpose and what is or isn't said in that description. Marketing claims will matter, too, but those will be scrutinized against the intended use – and should any discrepancies between these occur those will be treated as cancerous growth by default.

Fundamentally, the intended use is a promise from the medical-device manufacturer that their device can be safely used for the stated purpose under the stated conditions. Ultimately the intended use is ironed out at the end of the device development, but the better our draft of the intended use is at the start of development the smarter we are able to make use of our resources from then on. Understanding the intended use will greatly help us in planning and executing the usability engineering work needed for the device.

Above, in Section 2.3.1, we noted that occasionally the intended use might in fact be the use specification, or vice versa. Based on this reference, the IEC 62366-1 standard appears to softly expect that your intended use addresses, as appropriate, the following aspects expected of a use specification:

- Intended medical indication
- Intended patient population
- Intended body part or tissue type interacted with

- Intended use environment
- Operating principle

The intended use will share many elements with the use specification, and the two must at least be compatible with one another. In other words, if this was the beginning of the 19th century and we were setting out to build a railroad from New York we should at the very least know if we wanted to reach San Francisco or Key West at the end of our endeavors. Such a vague intended use would probably ultimately not do anyone any good, but it could be all we had at the start of the project. There is no requirement here that we know the ultimate goal – or the final intended use – from the start, but doing so will allow us to best plan the use of our resources and it will allow us to ensure that our intermediate work makes sense for the final outcome. See Section 10 for discussion of how to build from a concept to a releasable medical device with a final intended use.

4.1.2 User profiles

Rather surprisingly, the standard itself, IEC 62366-1, does not go into detail on how we should record user profiles so that they would be as practical as possible and address all the expected characteristics of the user. The guidance, IEC 62366-2, provides a little more detail by iterating the sort of properties we might be interested in users such as age, occupation, education, job responsibilities, and possible limitations (see Section 1.3). The guidance also points out that when creating user groups out of real user profiles we might want to bin these properties so that instead of unique values for properties such as age we would have meaningful value categories (e.g., child, adolescent, adult, senior). This is sound advice also when developing user profiles or personas. Table 4.2 shows an example of a user profile with some typical fields as well as a few additional fields we suggest facilitating observation of possible risk factors and dependencies between user profiles or user groups.

In the table most fields are self-explanatory, and all fields are entirely optional as the standard does not require any specific fields here. The fields we have included in the table, for the most part, do discuss aspects the standard cares about. Our novel contributions here over both the standard and its guidance include the following:

Table 4.2 Example user profile

Profile	*Registered staff nurse*	
Demographic details	Occupation	Clinical nurse specialist, staff nurse
	Job experience	1–5 years
	Age	Adult (21–60 years)
	Gender	Male/Female/Any
Education	Degree	A bachelor's degree or an associate degree in nursing (4 years), or less commonly a diploma from a nursing program given by a hospital or medical center (2–3 years). Education in, e.g., anatomy, physiology, social and behavioral sciences, communication, leadership, and critical thinking.
	Certifications	RN license.
	Skills	Specialist medical knowledge, healthcare environment, communication, leadership, and critical thinking.
	Training on this device	Not expected.
Impairments	Perception	None expected.
	Cognition	None expected.
	Dexterity	None expected.
Performance	Use environment and context	Hospital, medical center, or healthcare environment. Hospital ward context including ER and ICU.
	Frequency of use for the device	Infrequent (once per month)
	Attitude toward accompanying documentation	May read the initial training package (i.e., quick guide). May refer to user instruction during use of the device if time allows.
Interaction with the medical device	Promoting safe use	Medical training. Job ethics. High regard for patients.
	Detracting from safe use	At times hectic work environment. Distractions from other devices (e.g., alerts).
Relationship with other users	Dependent on	Doctors.
	Influence over	Patients.

- **Impairments**
  We suggest breaking down potential impairments into groups that can then be more systematically addressed during product development. No pronounced challenges regarding any of these subgroups of limitations shown here are to be expected with this user profile, but this may not be the case with some other user profile, and it may not be the case with every individual falling under this user profile. Examples of impairments might otherwise include colorblindness, bad eyesight, and hearing loss.
- **Performance**
  We suggest addressing the use context here under performance to look at both positive and negative factors of the environment and the user attributing to the performance of the work. The use of accompanying documentation, or any expected attitude or ability toward using such documentation, may be particularly interesting here as risk mitigation may occasionally rely on the use of such documentation.
- **Interaction with the medical device**
  We suggest looking at both positive and negative factors concerning the safe use of the device. Safety is the paramount goal of usability engineering for medical devices, and as such should be on our mind always.
- **Relationship with other users**
  We suggest identifying known links between the users and user groups in order to better facilitate analysis of the propagation of risk factors and the dependencies between the various user profiles and user groups. It is your responsibility to identify the users relevant to your device and address all these groups as appropriate.

The fields you end up selecting for your user profiles or user group definitions will no doubt vary based on your circumstances. No required fields are given by the standard or its guidance, but the above fields are a reasonable approximation of both what is typically expected and what might help you organize your expected users in a meaningful way. The choice here is yours.

4.1.3 Use environment

If a device is intended to only be used inside a hospital or at home, it may sound simple enough to define the environment as a cube of some dimensions and with running water or ample electricity. If, however, the device might also have to survive going through an MRI examination at the hospital or if that home user might take the device to the beach or the steam room with them this simplicity may be more elusive than we thought. Similarly, if a device is routinely carried through wind and rain by homecare staff, the environmental variables the device is subjected to will have greatly expanded from those expected in a laboratory setting. Depending on the device and its intended use, such use may be part of normal use or some form of misuse.

If, after a period of home use, a device is returned to the hospital without its batteries and the door to the battery compartment is missing, you can probably assume the patient has kids. If the returned device shows signs of having been disassembled entirely, you can perhaps assume the patient has a husband. And, if during a routine home visit, you notice that the husband has run his own electricity wiring from his wife's new oxygen concentrator and out the window, just so he could heat the car outside, that too may be quite normal but entirely unacceptable.

The point of the above real-life example, as colorful as they are, is to drive home the fact that not even a home is just one thing for everyone. All environments may introduce issues and once we move beyond a regulated environment consisting of professional users those issues may swing wide and far. Moving between environments, too, may compound issues and introduce new ones. Your unenviable job as a manufacturer is to think of all the relevant factors around your chosen use environments. Yet, the operative word here is "relevant".

Similar to user profiles, the IEC 62366-1 standard on usability engineering does not give us a map to navigate the particulars of describing the use environment. Annex H of the nonbinding guidance on the standard, IEC 62366-2, does give us a few pointers on the topic which we have considered in the following example. Table 4.3 shows an example of a use environment description with some typical fields and a few fields we suggest facilitating observation of possible risk factors and dependencies between user profiles, user groups, and use environments.

The fields you end up selecting for your use environment description will vary based on your circumstances, as always. No required fields are given by the standard or its guidance, but the above set is a reasonable approximation of both what is typically expected and what might help you organize your environments and care paths in a meaningful way. The choice here is yours.

Table 4.3 Example use environment description

Use environment	*Recovery ward*	
Defining characteristics	Purpose	Provide a controlled and monitored space for patients to recover after treatment. Medical professionals watch over the patients and are a hand to address any medical issues or emergencies.
	Size	Varies. The typical size is 5x7 meters. Size is expected to be sufficient for performing the stated purpose.
	Cleanliness	Hospital ward. No strict clean space requirements.
	Protective gear	Not typically expected unless called for by medical procedure.
	Climate	Typical hospital conditions. Room temperature (18–23C). Normal humidity (30–40%). Fresh air ventilation.
	Equipment	Electricity.
Ambient distractions	Visual (including lighting)	High-level lighting for standard care environments. 1–2 other patients in the same room and the medical devices in use by these patients (i.e., patient behavior and visual alerts from other medical devices). Sunlight may be present.
	Aural (including noise)	Quiet environment for standard care. 1–2 other patients in the same room and the medical devices in use by these patients (i.e., patient behavior and aural alerts from other medical devices). Note: due to emphasis on patient recovery, uninterrupted sleep is valued. Thus, audible alarms are not preferred and may be silenced on devices not viewed as essential.
	Physical	1–2 other patients in the same room and the medical devices in use by these patients (i.e., patient behavior).
	Other	Distractions caused by other devices used by patients or carried by staff.
Interaction with the medical device	Promoting safe use	Purpose-built care environment.
	Detracting from safe use	Other patients with other medical devices. Medical emergencies. Medical-device alerts from other devices.

Table 4.3 (Continued)

Use environment	*Recovery ward*	
Users present	Expected to be present	Patients.
	Often present	Doctors. Nurses.
	Intermittently present	Transporters. Visitors. Maintenance staff.
Related use environments	Similar environments	-
	Preceding environments	Operating room. Emergency room. Ambulatory care.
	Subsequent environments	Home care.

4.1.4 User tasks and use scenarios

After the above discussion of user profiles and use environments, this is no doubt getting to be old news, but IEC 62366-1 does not give us any template for recording or analyzing user tasks or use scenarios. The two terms are often used interchangeably, which is how we also use them here. The associated guidance, IEC 62366-2, does go into a little more detail to explain that task descriptions can be presented in, for example, narrative, tabular, or flowchart mode. This description is then noted to sometimes also address user perception (P), cognition (C), and actions (A) specifically (see Section 3.4.1).

Table 4.4 presents one model for recording the particulars of a user task. The example begins with contextual information and then proceeds to segment the task into steps and sub-steps as is relevant to each individual task.

In the example most fields are self-explanatory. The task title and overview are basic textual descriptions, as you'd expect, and the identification of users and use environments takes place as we have seen with the above examples of user profiles and use environments. Worth noting here is that we have opted to record the needs and benefits sought by each user group. Our intention is to facilitate a closer link with risk management and the purpose of each user's involvement in the task.

Recording the steps and refining these into sub-steps should be free to take place without any unnecessary distractions or extraneous details to

Table 4.4 Example user task description

<table>
<tr><td>Task analysis</td><td colspan="7">React to alarm for incorrect medication flow</td></tr>
<tr><td>Task</td><td colspan="7">The nurse notices an audible alarm sent off by a medical device indicating that the present intravenous medication flow has fallen outside set parameters. The nurse checks the present flow against the expected correct flow and makes appropriate changes. The nurse verifies that the change has been implemented.</td></tr>
<tr><td rowspan="2">Users and user needs/ benefits</td><td>Nurse</td><td colspan="6">Needs to keep the patient's medication inside predetermined parameters. The nurse can perform other tasks and attend to other patients while automation monitors the medication flow.</td></tr>
<tr><td>Patient</td><td colspan="6">The recovery of the patient needs the medication flow to stay within set, predetermined parameters. The device enables adherence to the limits in the care.</td></tr>
<tr><td rowspan="2">Use environments</td><td>Intended environment</td><td colspan="6">Hospital ward.</td></tr>
<tr><td>Foreseeable other environment(s)</td><td colspan="6">-</td></tr>
<tr><td rowspan="2">Risks involved</td><td>Risk IDs</td><td colspan="6">R1, R2, R3</td></tr>
<tr><td>Notes</td><td colspan="6">-</td></tr>
<tr><td colspan="3">Steps</td><td>P</td><td>C</td><td>A</td><td>Users</td><td>Risks</td></tr>
<tr><td rowspan="3">1</td><td colspan="7">Notice an audible alarm</td></tr>
<tr><td>1.1</td><td>Hear an audible alarm</td><td>x</td><td>-</td><td>-</td><td>Nurse</td><td>R1</td></tr>
<tr><td>1.2</td><td>Identify the correct device</td><td>-</td><td>x</td><td>x</td><td>Nurse</td><td>-</td></tr>
<tr><td rowspan="5">2</td><td colspan="7">Adjust medication flow</td></tr>
<tr><td>2.1</td><td>Locate the flow control</td><td>-</td><td>-</td><td>x</td><td>Nurse</td><td>-</td></tr>
<tr><td>2.2</td><td>Read the current setting</td><td>x</td><td>-</td><td>-</td><td>Nurse</td><td>R2</td></tr>
<tr><td>2.3</td><td>Determine the correct flow setting</td><td>-</td><td>x</td><td>-</td><td>Nurse</td><td>-</td></tr>
<tr><td>2.4</td><td>Adjust the setting</td><td>-</td><td>-</td><td>x</td><td>Nurse</td><td>-</td></tr>
<tr><td rowspan="4">3</td><td colspan="7">Verify medication flow</td></tr>
<tr><td>3.1</td><td>Locate the flow control</td><td>-</td><td>-</td><td>x</td><td>Nurse</td><td>-</td></tr>
<tr><td>3.2</td><td>Read the current setting</td><td>x</td><td>-</td><td>-</td><td>Nurse</td><td>R2</td></tr>
<tr><td>3.3</td><td>Determine the setting is correct</td><td>-</td><td>x</td><td>-</td><td>Nurse</td><td>R3</td></tr>
</table>

record. That's why we'd consider task analysis as a multi-pass activity where we first record the overall context (top part of the table) and then move to define the top steps (e.g., 1,2,3). Once this is done, we might move to sub-steps and add further details on the same row. The fields we suggest here for use with every sub-step consist of the following:

- **P (for Perception)**
  Tick this box if performing this sub-step may present issues for users with perception limitations (e.g., hearing, vision). You may decide to develop this field from a simple yes/no selection into some appropriate categorization as your circumstances dictate, or include such considerations in the body of the step description.
- **C (for Cognition)**
  Tick this box if performing this sub-step may present issues for users with cognitive limitations (e.g., elderly users, users with specific medical conditions). You may decide to develop this field from a simple yes/no selection into an appropriate categorization or description as your circumstances dictate.
- **A (for Action)**
  Tick this box if performing this step may present issues for users with physical limitations (e.g., mobility, dexterity). You may decide to develop this field from a simple yes/no selection into an appropriate categorization or description as your circumstances dictate.
- **Users**
  Here we take note of the users involved in this sub-step. If the same users/user groups conduct each step this may not matter, but in case the task has transient users coming and leaving the context somehow this will be highly interesting to the task analysis.
- **Risks**
  Enter IDs or descriptions of risks potentially involved in the conduction of this sub-step. You might also decide to flag sub-steps prone to use errors here instead of risks (see following paragraph). Addressing risks here in some practical way will ease your work later when you need to identify use scenarios related to hazards.

Instead of, or in addition to, risks you might also look at use errors. Use errors might either lead to or not lead to harms with controllable risks. The reason we opt to look at risks directly here is that we feel every single step we describe might involve a use error, and recording all of these here could hide the real issues under inconsequential noise. We, therefore, apply a modicum of restraint here and only record risks along with the steps when potential harm does appear to exist. Your choice between looking at use errors or at risks at this stage will perhaps depend on the risk profile and complexity of your device. Our approach here is in keeping with the purpose of task analysis in identifying use errors that can lead to harm, not just all possible use errors (some of which will be benign). Note that above we opted to look at benefits on a higher level already at the top of the form, while we record risks with steps. You could also decide to look at benefits along with each step, but we fear this will lead to unnecessary complexity in most cases.

Note that a record fashioned after the example shown here may also prove useful during usability evaluations if you are able to mark issues in, for example, perception, cognition, and action during the conduction of the tasks over the body of testing. This may give you actionable information on where in the set of tasks and the steps involved the issues lie, and how they should be fixed. Given the usually small test group sizes this suggestion may have us lean too far over the ship railing for our own good, but it may also lead to just the right intelligence for improving usability.

The guidance to the standard, IEC 62366-2, also voices its view of what we should record when observing users perform the tasks during some usability evaluation. These include the following:

- Was the user able to complete the task? How safely?
- What use errors, close calls, or use difficulties occurred? Why did these occur according to the user?
- Any remarks offered by the user on the use of the interface.

The example form given in Table 4.4 may be expanded to allow annotation of such properties during an actual usability evaluation.

4.1.5 Use specification

The expectations for preparing a use specification were discussed in Section 2.3.1, and these were noted to also apply to UOUP components in full in Section 2.5.1. In practice, developing a use specification should follow from the development of user profiles and use environment descriptions quite naturally. The required information here is in many ways a product of those deliverables, and a concrete statement of the who, what, and why of the device we are developing. Much of the information will not only instruct usability engineering work but also feed into product development work and the later regulatory submissions.

Table 4.5 shows an example use specification, written in a tabular form, covering the typically expected elements (in essence, the left-most column of the table). The table also adds contraindications (for use in later medical device submissions) and foreseeable extensions to some intended properties (for risk management purposes). The last item in the table, user tasks, is not required by the IEC 62366-1 standard here – although definition of tasks is at the core of the standard – but it is suggested by the accompanying guidance in IEC 62366-2.

Note that instead of redefining all these elements in the use specification, the specification may also refer to other source records as is practical. In particular, user profiles (see Section 4.1.2) may be developed to define the intended users and patient populations, and separate use environment

Table 4.5 Example use specification

Use specification		
Intended use	{Description or link to the intended use, preferably both}	
Indications	Indications	{...}
	Contraindications	{...}
Patient populations	Intended populations	{e.g., age group, weight range, and health}
	Foreseeable other populations	{...}
Body part or tissue type interacted with	Intended	{...}
	Foreseeable	{...}
Use environment	Intended environment	{e.g., sterility, reuse, home use, protective gear}
	Foreseeable other environment(s)	{...}
Operating principle	{e.g., laser, stainless steel blade, intravenous tube}	
User tasks	{anticipated user tasks or needs}	

descriptions may be used to address environments in a controlled, reusable manner (see Section 4.1.3).

Note also that instead of a tabular format, the use specification you write may also be a textual description discussing some or all the above characteristics. Your use specification, or a summary of it, may be your intended use, but at the very least the two must be in synch. We feel that a tabular format will allow you to sort the information better and also maintain whatever links between entities you wish to make use of in the usability engineering activities. The choice here, too, is yours.

4.1.6 User interface specification

The essential requirements (see Section 2.3.6) for a user interface specification are that it matches the use specification just developed and that it observes any identified use errors and hazard-related use scenarios.

First and foremost, the specification must include testable technical requirements including any risk controls needed. IEC 62366-2 Clause 15.3.2 provides a few such example requirements calling for the use of meaningful screen titles (or breadcrumbs), progress bars, particular font sizes, and even dynamic elements showing the software has not halted. The examples here are simple, but illustrative of the kind of requirements we might have in our specification. Additionally, the specification should make it clear whether accompanying documentation or training is needed to use the interface.

Rather helpfully the guidance document points out that requirements can be sourced from, for example, general usability engineering literature and the manufacturer's own accumulated collection of customer requests. Such knowledge may often take the form of silent knowledge that travels between projects between the ears of the experts, but it can also be developed into a practical database of requirements to draw from. A written form is, of course, required for use in the documentation.

As with any product development document, the contents of the specification should be kept up to date as work progresses – the specification is hopefully fully developed ahead of time, but if changes do take place these should be reflected in the documentation. But beware: don't just update files with timestamps to show progress has been made. That truly helps no one and only increases bureaucracy and uncertainty (see Juuso 2022).

The nonbinding guidance document pulls its socks up here and asserts that also requirements for the accompanying documentation and training should be developed here. The language is much more forceful than the standard itself allows – the standard only asked for the necessity of documentation and training to be considered – so treat this expectation with a grain of salt. If you do have requirements for documentation and training by all means document these here, but don't wrap the anchor around your legs and feel that you have to develop as verbose a set of requirements for your accompanying documents and training as you have for your medical device. Having said that, a body of standardized requirements for documentation and training may prove reusable from one project to the next. Our objection here is only with the loose escalation of requirements sought by the nonbinding guidance, not with the merit of such expectations.

The guidance also suggests that training materials are developed when now working on the user interface specification. This may indeed be possible, perhaps even beneficial, to considering how each suggested requirement later manifests itself to the user (e.g., how much explanation a requirement may need once presented to the user), but you may also be polishing a layer of concrete you will rip up several times before you are actually done with the final interface. We would thus look ahead but be wary of trying to jump ahead.

Table 4.6 sketches out what a UI requirement might look like as an individual entry taken from a database of requirements specified for a single interface. In addition to what is shown in the table you may want to include references to the necessary system requirements or software requirements as instructed by IEC 62304 (see Juuso & Pöyhönen 2024). Note that an ID or an approval status is not mentioned by the standard here, but these are almost certain to be needed in order to keep the records in order over any period of time.

Table 4.6 Example UI requirement entry

User interface specification	
ID	{Unique ID}
Requirement	{Textual description}
Documentation expected	{yes/no}
Training expected	{yes/no}
Status	{Approved / Draft / etc.}

Maintaining such requirements as a database instead of a burgeoning text document will seem tempting in many cases. Such a setup may make maintaining and even reusing requirements much easier and safer than performing copy-paste repeatedly. As long as ISO 13485 requirements on record controls are met (see Juuso 2022) and the appropriate reviews can be conducted (and effortlessly re-accessed thereafter) across the body of requirements stored in a database (see Juuso & Pöyhönen 2024), there is no real reason to require documents.

4.1.7 User interface evaluation plan

We discussed the requirements for the user interface evaluation plan in detail in Section 2.3.7. The guidance document, IEC 62366-2, goes over the same ground and points out that typical project management details are beneficial here for ensuring our schedule holds and deliverables, such as prototypes, are ready for testing or whatever other synchronization is needed between product development activities. The discussion offered by the guidance here is appropriate and rings true, but it is also very much a repeat of everything already said by it and the standard earlier. The expectation here is that we make the plans we need for performing the formative and the summative assessment work, and in particular know the selection scheme for the use scenarios to include in the summative evaluation.

The key set of information to cover in the evaluation plan could also be described as the test protocol. The protocol might, for example, consist of the following information:

- Purpose of the evaluation
- Evaluation method (e.g., expert interviews, test groups)
- Data collection and analysis methods (including forms used)
- Identification of the tested device (or its parts)
- User tasks to cover
- Supporting materials needed (if any)
- Associated training needed (if any)
- Associated accompanying documents (if any)
- Test environment
- Test participants (i.e., the subjects)
- Test personnel (i.e., the testers)
- Any supporting material used by the testers
- Approval of the protocol

The list above is not exhaustive. As always, what matters is that we take note of all the material used and record the evaluation as closely as necessary for possible later review and repetition if needed. The standard refrains from making any micromanaging expectations here, but the standard scientific process and expectations of repeatability will always apply here. Your plan, report, and reality should match here – and if they don't, you should act accordingly.

Perhaps the most practical way of looking at the user interface evaluation plan is to think about it in terms of the parts of the whole device you want to address through your usability activities – and whether you want to address these parts through prototypes or only via the finished representative product. This will give you a roadmap of formative evaluations to run and also link these evaluations to the interface mockups or maturing versions of the device you will ultimately want to create. Afterall, some evaluations can be run using post-it notes while others will require some code to also exist behind the pretty interface to serve up appropriate responses to interactions. These will then all culminate in the ultimate summative evaluation at the end of your roadmap. The information on this roadmap is all of the type you would expect to see in a typical project plan (e.g., schedules, phases, people, dates), but the purpose of it all is uniquely tied to the maturing usability of your device.

4.1.8 Usability evaluation reports

The standard itself, IEC 62366-1, discusses a single plan for usability evaluations (see Section 4.1.7) which is to address both formative evaluations and summative evaluations as is appropriate. The standard then expects to see formative evaluation reports and a summative evaluation report as the plan laid out. No separate capstone report tying these activities together is expected by the standard, but you may use a part of your final summative evaluation report to also look back on earlier formative work as is appropriate for the conclusions you draw.

However, the nonbinding guidance to the standard, IEC 62366-2, does expect a final capstone report summing up all formative and summative work. The sections the guidance envisions for the usability evaluation report are:

- Executive summary
- Use specification

- User interface specification
- Known use problems
- Description of selected hazard-related use scenarios
- Formative evaluations
- Summative evaluation
- Conclusion

The guidance suggests we cover the above topics by writing appropriate summaries and descriptions here in the capstone report. Alternatively, you might choose to reference appropriate documentation already created for some of the topics. The executive summary and conclusion are naturally the product of new reflection and assessment made here at the end, but you will have already addressed many of the other topics in your earlier records such as the final summative evaluation report. Regurgitating information helps no one and confuses some, but you should, of course, ensure that you review all pertinent information and only then draw conclusions.

We dislike the idea of yet another report. Our approach would be to instead position the final summative evaluation report as the ultimate report on your usability evaluation activities. This must exist in any case, and if no formative evaluations were planned it will also be the only usability evaluation report required to exist. The final summative evaluation report will also already have access to all the information listed above by the guidance, including any conducted formative evaluations and their relation to the summative evaluation. We feel that in an effort to be helpful the guidance here ends up generating unnecessary paperwork that adds nothing new to the story but is nonetheless something additional to maintain up to date later on as we revisit usability engineering activities through, for example, product changes and post-market surveillance. We would not listen to the guidance here, the standard definitely does not, so think twice about creating a capstone usability evaluation report.

For the expectations of formative evaluation reports and the summative evaluation report see Section 3.12 and 3.13, respectively.

4.2 Risk management documentation

As the focus of medical-device usability is on safe use instead of snazzy designs and accounting of the minutes purportedly saved, risk management is in many ways at the core of the discipline. Risk management is the creamy filling inside the truffle of usability, except that sometimes this filling can be made of salty licorice instead. The IEC 62366-1 standard looks towards the industry standard ISO 14971 on risk management for instruction of the information to keep in the identification, analysis, control, and overall management of risks. It does, however, point out several areas of joint interest, as we will cover next. In this section we won't provide example formats for addressing the information as we feel your risk management activities already provide enough instruction on the topics covered. We will, however, point out the special areas where usability engineering can prove useful in developing that information.

4.2.1 Hazards and hazardous situations

The identification of hazards and hazardous situations is square one for medical-device risk management, and it may also be that for usability engineering (although something touching on appeal to users, marketing needs, past experiences, and the organization's product portfolio may predate that square). The emphasis in usability is naturally in the role usability and particularly use errors may play in making the devices safe.

We discussed the requirements for identifying hazards and hazardous situations back in Section 2.3.3. As for recording that information, we feel that both these items are columns in your overall risk management table or entries in your risk management database should you have something more purpose built than a spreadsheet to work with. For us, a spreadsheet nicely illustrates the needs here, and ISO 14971 perfectly captures the information to be recorded. Hazards and hazardous situations will not be unique to usability engineering, but rather they will be the same hazards and hazardous situations that you deal with in the overall risk management of your device (although see Section 9.2 for discussion of how the same risks can be addressed through many different points of view).

Usability engineering should prove itself useful to understanding the use, users, and use environments of your device from the very beginning of your design and development activities, but unless your risk management activities are somehow missing hazards and hazardous situations where usability may play a role, it probably doesn't have many hazards or hazardous situations to add to the list. Situations vary, of course, but mostly the usability standard and its related guidance are happy following the lead of ISO 14971 here.

4.2.2 Hazard-related use scenarios

As discussed above, hazards and hazardous situations are perhaps more fodder for ISO 14971 risk management than IEC 62366-1 usability engineering, but where the two really start to find a common tune is in the identification of hazard-related use scenarios.

Traditionally, risk management is concerned with all the things that could go wrong in attempting to perform some positive action. Today increasing interest is placed on understanding the benefits associated with those positive actions (see Section 6), but the emphasis is still on thwarting the negative actions. For the optimal use of resources and controlling risks this is as it should be. In usability engineering the scope is decidedly wider: here we look at not just what could go wrong, but also what can go right, and what else may happen. Even for medical devices, where the focus is on safety, the activities may encompass both correct use and use errors (i.e., normal use). Thus, risk management looks at controlling negative events while usability engineering mostly looks at enabling positive events. This is a marriage made in heaven as long as our resources allow the activities and the goal of creating a safe medical device is shared by both sides.

In terms of documenting hazard-related use scenarios, we actually already addressed this in Section 4.1.4 while discussing user tasks. There we linked steps in the task to risks defined in our risk management. As we already did this, it should now just be a matter of pulling up all use scenarios (i.e. user tasks) and filtering out those without a risk associated with them. In practice, the analysis of whether risks are involved and which risks are grave enough to warrant inclusion in our summative assessment may require more thought and even iterative work. You may also want to perform some summative analysis on how similar or dissimilar the scenarios are that have associated risks and if there could be some benefit to implementing uniform risk control measures. In principle, to just record hazard-related use scenarios we would lean on ISO 14971 and observe the links to our usability documents. It is possible, but perhaps not always practical to go into very much more detail here.

4.2.3 Use errors

The concept of a use error is deceptively simple: IEC 62366-1 Clause 3.21 defines use errors as user action (or lack thereof) which leads to a result not intended by the manufacturer or expected by the user. A malfunction of the medical device, or an unexpected psychological response from the patient, are potential issues but they are not considered use errors. The inability of the user to complete a task might, however, be considered a use error if the manufacturer should have done something differently. The link between a use error leading to harm is thus intuitively similar to the link between a risk and the associated harm, but use errors and risks are not the same. As discussed in Section 1.6, a use error might even be benign or positive in nature when a risk is always deemed negative in the context of medical devices. The question thus presents itself: how do we document use errors?

In risk management, a use error might show up as an initiating cause or some contributing factor to a risk analyzed on multiple sequential levels. In usability engineering, use errors should be recorded as arising out of the use scenarios where they have been observed or where they are thought possible. Use errors and close calls will be observed in formative and summative evaluations, and should thus show up in their reports and then somehow instruct risk management for the device in question. After a device has been released, any arising issues will be analyzed in the light of the intended use, instructions for use (IFU), and any marketing message given by the manufacturer. All of these contexts should link to the same risk management and usability engineering file of the device somehow. How exactly this is accomplished is, to a large extent, left to the manufacturer to organize and the conformity assessment bodies to assess.

To further complicate matters, use errors effectively exist on two planes: 1) a theoretical plane constructed before any usability evaluations are made, and 2) a practical plane of use errors (including close calls) observed during usability evaluations and real use after the release of the device. This might even be considered three planes, but for us the distinction of merit here is between something anticipated and something encountered. Both should show up in the usability engineering file and the risk management file as appropriate. We might therefore suggest the following for documenting use errors:

- Document use errors as part of risk management from the very beginning of design and development activities. When you encounter user errors or risks as part of usability activities (e.g. developing the use scenarios), enter these in risk management and leave links as suggested in Section 4.1.4.
- Document use errors in your formative and summative reports. Ensure that the conclusions made in these are escalated to your risk management activities. Use this information to collect data on what use errors are known to have occurred in practice and with what effects.
- Later as new use errors and risks are detected during post-market activities, analyze these through usability engineering considerations and modify your risk management file as is appropriate.
- When you later revise the formative or summative assessments, proceed as before with the appropriate effects on the risk management file.

The above is our recipe for a lean approach to coupling usability engineering and risk management activities in a useful way. Your approach may look different, and it may be more complex depending on your circumstances. There is no real right or wrong way of addressing the overlap of information from these two domains as long as the model you adopt works for you.

Where we would suggest you apply some restraint is in documenting the benign use errors – if something is not related to a harm or a risk, and it doesn't help you uncover new benefits either, it may just be noise chipping away at your resources both now in generating the documentation and later in analyzing and maintaining that information. The IEC 62366-1 standard enables the addressing of benign and positive events, but the extent you do so is up to you.

Similarly, if you observe the issue during usability assessments that the device was dropped on the foot of the user and broke a few toes, but then discover that this was because the technician accidentally first dropped the cigarette they were smoking while carrying the device, the analysis and ramifications of the event may be short lived. You should record the event and make the necessary conclusions, but to record all such events in minute detail might be counterproductive and even detrimental to the overall analysis you hope to develop. The magnification glass should be pointed where it matters, and not just at what is interesting. The point of the exercise is to make the device safer, not to force the technician to quit smoking (although he probably should not have been smoking while carrying the device).

4.2.4 Safety-related user interface characteristics

We already discussed safety-related user interface characteristics at length in Section 2.3.2. Our discussion there began with these characteristics and how understanding them could help us analyze hazards, hazardous situations, hazard-related use scenarios, and usability evaluations better. Here we chose to reverse the order so that our discussion of documenting the usability engineering activities end with the characteristics of a safe user interface we had learned. We thus bookended the discussion of usability documentation with the characteristics. Our motivation for doing so was to drive home the fact that usability activities increase our knowledge of the users, use environments of our device and this information leads to not only improved risk management but also a better, safe device. The latter means that we should thus here understand what characteristics signify safe usability and how those affect the requirements of our device as it progresses through our design and development activities.

You may document the safety-related user interface characteristics as a report or, for example, by categorizing some of your product requirements as stemming out of usability considerations. Such requirements should, of course, be part of the considerations in reporting your formative and summative usability evaluations. Our suggestion here is that you make sure you close the loop from the theoretical to the real world, and from the design to the observed evaluations. Doing so will be easier if you have some succinct reports and nifty database system to work with, but realize that it will never be a state-of-the-art solution to have a fluid set of characteristics with no assessment of the whole set and no follow up. Thus, a database accompanied by meaningful status reports at practical intervals and steps throughout the design and development work will probably make the most sense. The choice, once again, is up to you.

4.3 Other development documentation

We have already covered the typical usability engineering documents above, but in addition to these, a body of other development documentation will also exist that is in one way or another associated with the usability engineering documentation discussed above. These may be development documents referenced from usability documents, development documents referencing usability documents, or merely documents existing on the same timeline with implications on the usability engineering side of activities.

We addressed such development documentation in detail in our two previous books (Juuso 2022, Juuso & Pöyhönen 2024), and will again briefly return to the topic in Section 10, so here the only point we want to make is that referencing such documentation instead of copying it over to some special usability engineering folder will make the most sense. This should go without saying, but it doesn't always work out that way in real life, so we decided to offer this note here. All the documentation you develop around your device, in whatever activity, will need to be cross-referenced as appropriate and traversable in some practical way to prove your actions and reasoning in later reviews, so this is nothing new really. Keep it simple.

4.4 Accompanying documentation

Accompanying documentation might technically not be a part of the core of your usability engineering file, but it is quite quickly implicated in any audit of usability engineering as the outcome of usability work during development may instruct the development of this documentation. If accompanying documentation is used as a risk control, if it contains necessary information for safety, or if you somehow claim that issues discovered during usability evaluations were addressed via such documentation, it is necessarily a part of the documentation looked at as part of the usability engineering file.

IEC 62366-1 Clause 3.2 defines the term "accompanying documentation" as materials accompanying a medical device and intended to provide the user with information for the installation, use, maintenance, decommissioning, and disposal of a device, particularly regarding its safe use. Note that the last two items were added by the 2020 amendment. Specific types of documentation falling into this category might, for example, include the following:

- Instructions for use
- Technical description
- A quick reference guide
- Instructions for delivery, storage, installation, maintenance, repair, and decommission

- Release notes
- List of residual anomalies
- Warnings, icons, labels
- Training materials
- Marketing materials

The standard is also aware that some devices that can be safely used without documentation may not need accompanying documentation in some jurisdictions. At the same time, translated versions of many user-facing documents (such as manuals) may be needed, and their usability evaluation may be addressed via different mechanisms (see Section 1.14). See Juuso & Pöyhönen 2024 for more discussion of accompanying documentation.

4.5 Post-production documentation

The "evert after" -part of the device lifecycle is receiving increasing attention from regulators and standards bodies. Not only is usability-related material generated as part of investigating complaints and developing product updates, but also a wealth of other post-market surveillance and post-market clinical follow-up information is required to be developed which may then also be of interest to your usability engineering activities. You probably shouldn't go overboard requiring that such information is periodically reviewed to assess usability considerations, but you should ensure that usability considerations are addressed where they are appropriate in your other QMS activities.

The prime suspects of documentation to consider are:

- Periodic safety update reports (PSUR)
- Post-market clinical-follow-up documentation
- Periodic or revised risk management reports
- Any new usability engineering reports
- Change management documentation
- Problem resolution documentation

Of the above, particularly the periodic safety update reports (PSUR) should prove useful in assessing the overall state of your products. This report will be based on your sales and usage data, any encountered incidents and complaints, and the feedback you may have received. Starting with such a report already created in another process and summarizing much of the information you might be interested in from a usability point of view will save you significant effort in collating information and analyzing where you want to spend the little resources you may have available in running post-market usability activities. The information you get will probably only be about as good as the usability engineering skills of the people involved in creating those reports in the first place so it will be worth your while to ensure usability awareness propagates through your organization in much the same way as awareness about ISO 13485 quality management is required to.

See Section 9 for more discussion of how usability engineering connects with the other QMS processes.

Chapter 5

Clinical Environment

Science fiction and period dramas hold special appeal because, for one thing, they allow us to immerse ourselves in foreign environments with customs, rules, and expectations markedly different from those we see around us every day. That sort of inquisitive adventurous spirit is what we hope to stoke here in this section as we promote a need to understand the clinical context from the side of manufacturing, and vice versa. Diving into these new environments can be a scary proposition, but hopefully one with a cheerier outcome than that illustrated by Figure 5.1.

This section attempts to apply just enough sandpaper to the fence posts between the manufacturing world and the clinical environment so that people on either side are not afraid of at least talking across the border. The device will be assembled in the former but put to the test of time in the latter. It therefore only makes great sense to ensure that nothing is unnecessarily misinterpreted or lost in the translation in between, regardless of which direction we move in. Our objective here is not to build a complex model of how such interactions could take place from here to eternity, or take into account each and every specific individual context but rather to perhaps write down 3% of such a grand unified theory – just enough to smooth over the first such cross-border interactions. After that the next steps will usually be self-evident. If you already frequently move between the two domains there won't be anything new for you here, but if you sit firmly on either side of the divide and wonder about the other side we hope that the anecdotes and real-world advice we have assembled in this short section will get you aligned into an appropriate entry trajectory and position you to better understand what you will see once landing – or

DOI: 10.4324/9781003645702-5

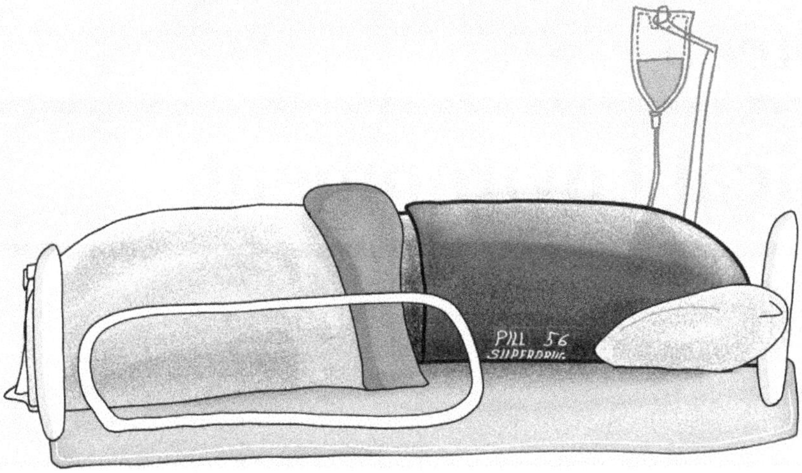

Figure 5.1 Here rests the devil in the details, vanquished by the will to try.

what you'll see when you open that air lock from the outside in. Images from the opening scenes of the original 'Planet of the Apes' rattle around in our minds as we write this section, but only to the extent of landing in some fantastical new environment. The discussion here is a little above and beyond a discussion of the IEC 62366-1 standard, but at the same time we feel that it is absolutely warranted as we seek to improve the level to which manufacturers understand their users, and vice versa.

The term 'Renaissance Man', besides a lesser-known Danny DeVito film, refers to a human whose knowledge is so broad as to encompass everything, or at the very least everything of relevance in some particular setting. An 'Uomo Universale', if you will. The most famous specimen in this regard is Leonardo da Vinci (1452–1519), who was not only a painter and a sculptor, but also as scientist, theorist, and an engineer. His knowledge of human anatomy and interest in making sense out of the human musculoskeletal system also paid dividends in his paintings and sculptures of the human body. In his day it may still have been possible for a single person to possess a command of the whole scientific knowledge, but today that is no longer possible for anyone this side of generative AI. What matters is that we are able to discuss matters across disciplines and collaborate on solving identified issues.

We have both spent our careers in one interdisciplinary context or another. This includes research, manufacturing, and conformity assessment, but also significant time spent at the hospital wearing various hats and visitor passes. We hope to distill some wisdom from this experience and provide a little insight into both sides of the equation, so that each side can perhaps understand one another a little better should this be your first time contemplating a trip across the border.

Years ago, we had a joint research project between our university and a large regional biobank run at a university hospital. The purpose of the project was to assist the biobank in developing biomedical analysis software, consolidate data from multiple regional biobanks together, and also assist in devising a working business model for the biobank. On the first day of the project, our away team from the university was summoned to a meeting at the hospital where also a good number of the doctors and staff from the biobank were present. In all, close to 30 staff members packed into a large oblong meeting room. The visual of the meeting was that of the opening scene in the Godfather, only instead of one Marlon Brando sitting in a room backlit with sunlight squeezing in through window shutters, there were 15-plus brandos, men and women. The talk was not of asking favors on anyone's wedding day, but certainly both sides were trying to figure out who the other one was and what they wanted to get out of the marriage. The group of doctors turned out to be great professionals in their field, courteous and friendly in nature, and packed with wicked great senses of humor: one senior pathologist joked that we all need to work hard for the next 30 years so that we get to enjoy the fruit of our labor during the last 30 or so days of our lives. He meant that the work we were about to embark on would improve the care of patients, and perhaps even lead to new treatments, but at the very least the benefits could be felt by each of us when returning to the hospital at the time many people do, very close to the end of their lives.

Setting off on a convivial footing, we started our deep dive into this new healthcare setting, which was more academic than clinical, but separated from hospital wards by only a few sets of doors. Many of the doctors traversed both sides of those doors, but in general not even all doctors are at home in both academia and the clinical world. By working at the biobank, we got an incredible introduction into the world of medicine, but we perhaps got to wade in through the shallow end of the pool given our university backgrounds. Nonetheless we learned that we had similar, compatible goals across the table but perhaps intuitively worded these differently. We also observed some hierarchies and levels of seriousness that maybe weren't as pronounced at the non-medical university campus, but none of this hindered us from working together. At certain meetings it was necessary to be coached a little on the appropriate etiquette beforehand, and in others it felt like we were meeting with some high-level foreign diplomat and negotiating a deal that our country needed to survive. All of this was new and fascinating, foreign in parts but readily solvable using the vocabulary we knew.

The biggest lesson we drew from this immersion is that you should not go into a new context bullishly, but instead listen to the folks you encounter, link what you hear with what you know by training, or otherwise, and use the map to translate over any hurdles you come across. That may be as consoling as asking a fresh recruit to remember their training when thrown into war, as happens in several Hollywood war films, but it also happens to be somewhat true. Gradually you will notice that you are understanding more of what you are seeing, and you are able to discuss it all constructively. Through the process you will also start to understand more than just a little bit about the other side of the equation, how people in that context work, what they value, and how you can work together.

Another anecdote we'd like to offer here is also a story of diving into unfamiliar waters. Years ago, academic collaboration took us to Japan for the first time. Jet lag from the long flight to Osaka had taken its toll so we were having some trouble finding a restaurant marked on the small map we had gotten from our hotel. Not being able to find the right place and not making much sense out of the Japanese store fronts we asked a passer-by. This was years before Google Maps and e-SIMs. They took the map, looked at it, ran to the next street corner and the one after it, and then returned with someone else and nodded profusely at a nearby door handing us the map. We were speechless at the display of absolute courteousness and only knew that we felt welcome in a foreign land which we could easily love just based on this one encounter. Later we learned, through first-hand experiences, that in Japan you got on busses through the backdoor and paid on your way out by placing coins on a small conveyor belt – and that if you press the "English" button on a subway vending machine enough times without effect a human being may pop out from small hatch next to the machine and know what you wanted without any dialogue. These events were just the tip of the iceberg of our experiences traveling between Osaka, Kyoto, and Tokyo in Japan, but a great example of how you should be ready to solve familiar issues in completely novel, even totally opposite ways and yet come up with perfectly practical and perhaps even superior ways of accomplishing tasks.

That is the mindset we have since tried to take when diving into new environments and contexts. If something you encounter doesn't immediately make sense to you, particularly if you didn't see it coming, it doesn't mean that it should necessarily be changed. Not having ever been to Kansas, yet, we don't really feel we have permission to use the saying that we aren't in Kansas anymore, but that is a good stance to take here if you now want to dive into the clinical context for the first time. The same also applies if the clinical setting is your home but you now want to talk with medical-device companies in one context or another.

One last anecdote on being thrown into new, strange, and wonderful environments is that our travels once took us to a digital humanities conference in Michigan. Being engineers and this being our first dip into the humanities, we especially enjoyed one evening mixer organized at a splendid old-world mansion. Wandering from one conversation to the next with a glass of wine in hand you never knew in advance how a conversation might go. Some participants had experience with software development, and they instantly felt like colleagues although our homes were spread across the US, Europe, and Asia geographically. With in some other encounters the conversation found a diving board to the deep end of the humanities scholarship discussing the works of Shakespeare, Samuel Beckett, Charles Dickens, or some other great name we should have been immediately familiar with if we weren't impostors coming in from a completely different background. This experience too was an eye-opening challenge that ultimately turned out to be life-affirming. Out of this experience came a profound appreciation for the need to listen, understand, and map the terms and concepts given to us with those we knew from before. The terminology and even the concepts at this mixer were different than we were used to, but the motivations, goals, and even the broad strokes of research workflows were the same. We can talk about the same things with completely different terms, but once you crack the code, we can all move ahead together towards better outcomes. In other words, at the beginning of such encounters you should probably expect some element of talking past each other, or some parts of the message getting lost in the translation, but after a while you will have created the necessary lookup table if you cared to listen closely enough. In the years since we have found this to also be true on the interface between medicine and engineering, academia and industry, too. We would even venture a guess that this same concept would also apply if Martians decided to land here tomorrow.

With those anecdotes, each based on real life, we hope we have set the scene for how to approach a first gathering of folks from the manufacturing side and the clinical environment. Both of course mean multiple different things, but if the staff of a manufacturer still need help talking among themselves, or if the clinical specialists do, that's perhaps a bigger concern and a sign that the time is not yet ripe for climbing across the fence. The tools to use even there are perhaps the same.

In the following we wish to offer a few brief remarks on the sort of things both sides should know about each other, and the topics which may become spheres of common interest. We present these remarks with certain knowledge that they don't provide a comprehensive guide to such encounters, but we feel that a brief discussion along with the IEC 62366-1 standard on usability engineering is more than warranted.

5.1 Spheres of common interest

Manufacturers of medical devices have a built-in need to converse with medical professionals to discuss the solutions they offer, sell products, and hopefully develop ever-better solutions to improve medical outcomes. Medical professionals, too, want to be aware of the best treatments, best tools, and best solutions they could employ as part of their everyday work in the clinical context. The notion of what exactly makes something "the best" may differ based on the context and the party making the judgment. The following lists a few competing definitions for what best-in-class actually means.

- **Performance**
  The better device achieves its purpose with unprecedented speed, precision, recall, or efficiency in terms of the resources used (e.g., time required from the nurses and doctors operating the device). The improvement here could be theoretically proven or the claims backed by evidence from some relevant studies.
- **Safety**
  The better device has a more convincing safety track record than its competitors. The device is harder to use in some unsafe manner, or it may come with fewer or reduced side effects.
- **Fit into existing care flows**
  The use of the device doesn't require changes to established ways of doing things and can be used by current staff members without

extensive training. An example here might be a device that can already be applied in the ambulance on the way to a hospital, or a device that doesn't require 10 new steps to go from the old point A to the old point B.

- **Wider applicability to relevant indications**
  The better device overcomes some of the limitations of competing devices. An example here might be an ECG patch that doesn't need to be taken off a patient when going into an MRI exam.
- **Pleasing or intuitive usability leading to reduced training need**
  Not only is the better device safer to use, but it is also more rewarding to use than the alternatives. The device doesn't require renewed training or frequent use of manuals each time the device is needed. The simple fact of coming with less irritation and annoying hangups is a case for good usability, but to demonstrate this good word of mouth, testimonials, and possibly a period of test use is needed.
- **Lower price**
  Purchasing or using a device doesn't come with scolding from the head of the department or some form of buyer's remorse. Price is, of course, a complex equation that may come down to many factors including reimbursement, patient vouchers, annual costs, sunk costs, costs for single-use accessories, reprocessing, and a wide range of other considerations affecting the total cost of ownership and use. A device costing less than its competitors will more often than not have a competitive advantage over its competitors.
- **Lower intimidation factor**
  The design of the device helps put the patient and the user at ease so that some procedure is easier to perform. This is why children's wards are often painted in bright colors and decorated with the images of familiar cartoon characters – or why a dentist will probably administer local anesthesia before drilling into a tooth. A device housed in a sleek white plastic case instead of looking like it was built by the janitors the night before will be better in terms of instilling confidence in its users. The flipside of this may be that a technique or a device taught at the medical school as part of their training may feel less intimidating to the medical professional than something newly introduced.
- **Improved patient retention**
  A crucial factor in the success of many medical treatments – be those pharmacological, device-assisted, or simply regiment-based – is patient adherence to the prescribed course. Taking the full course of antibiotics

instead of stopping on a whim once you feel better is notoriously important. Similarly, if the use of some device is easy to forget or hard not to hate its use may be dropped as quickly as a gym membership in February every year. A device that patients or other users like to use will be better from a retention point of view.

- **The X factor**
  The coolest, most appealing, most tantalizing, or simply the device subjectively most desirable might also win the crown of the best in show. For example, a toothbrush that uses generative AI might today be the hottest product out there even if we didn't quite know why putting AI in a toothbrush would be a smart move. The device might not meet any of the above characteristics, but it could still be the best because of some property X. Advertising, too, might win patients over and nudge them into requesting a specific product from their doctors.

Looking at the above list it is clear that some yardsticks might be more appealing to the manufacturers than the medical professionals, and vice versa. A few yardsticks, however, are topics of joint interest (e.g., performance and safety). The list is not exhaustive, but it is illustrative on the different trajectories we might all follow in approaching a discussion across the fence line. There are rules and conventions for how such interaction takes place, and perhaps even forms to fill in with the intention of weeding out undue influence across this interface.

Medical professionals may feel uneasy with the commercial setting this conversation may take place in – and the manufacturer too may feel like they use up unjustifiable amounts of resources just to perhaps participate in this conversation or to put together all the formal documents asked of them just to participate in a bid. This is all accepted as the way the world works, and the cost of doing business, but it is not exactly optimal for efficiently developing and selling medical devices, or getting the best devices for the best prices with reliable long-term support. All the overhead and unrequited work will be paid by someone sometimes, and some of the unsubstantiated promises made during the heat of pursuing sales may hurt someone sometimes. There is no magical Harry Potter spell to fix the inefficiencies here but understanding both sides of the table a little will perhaps help negotiate a good solution for going forward. It will lead to the right solution being assigned to the right problem, and ultimately the right products being developed and these products working as expected by the users.

5.2 Existing products to meet existing needs

Perhaps the classic case for a manufacturer venturing out into the clinical environment is to sell their device. The Will Smith film "The Pursuit of Happyness" (note that the spelling here is intentional) is based on a true story and provides ample examples of what it might look like when a salesman tries to push a medical device out of the boot of his car. Generating interest, accepting a trial use period, and hoping to get bucks on his bottom line is the name of the game here.

From the manufacturer's point of view the ultimate win here would be a long-term relationship leading to vast, accumulating sales of the devices they make in any case – that is without further specialization or development taking place. The invoices they get to send out should be large, steady, and ever-increasing, and the number of issues, complaints, and adverse events encountered non-existent. From the user's point of view the ultimate win would perhaps be to get the best solution to the issue they had, with no hidden costs or shortcomings now or over the period of use despite changing needs on their part. The price to pay should, of course, be as low as possible and fit into the budget, budget type, and budgetary year as relevant to them. The manufacturer will gain some bragging rights from getting their device into trial use or even real use at a prestigious institution. Similarly, the clinical institution will get bragging rights by adopting cutting-edge technology that their potential customers, their patients and possibly payors, will be wowed by. It is a win-win situation, no doubt, but one where traction has to be apparent after the initial infatuation.

From the manufacturer's point of view a miserable failure might be to spend vast amounts of unpaid work on developing an offer that leads to losing out to a competitor, or getting an offer accepted but not being able to deliver on the offer. From the user's point of view a loss would be accepting a bad offer leading to issues to fix, but also accepting a bid later challenged by their own organization or some competing manufacturer in some way. Loss of prestige or reputation following a failed trial could also hurt both sides, but perhaps more so the manufacturer.

In the real world the different types of payors, bids, purchasing organizations, advocacy groups, and even the types of manufacturers and manufacturer offerings are so diverse that we can't claim to offer any magical recipe to streamlining the interface here. Nonetheless, we feel that stating a few of the big strokes is beneficial for striking up discussion in the context of medical-device usability engineering.

5.3 Building clinical evidence

No company is set up to fulfill just the needs of one clinical customer. For that reason, much of the conversation may be focused on building up evidence for the product working well and meeting the expectations in pre-clinical or clinical use. Clinical evaluation and clinical investigation are topics for regulations, for ISO 14155 on the clinical investigation of medical devices for human subjects, for accumulated clinical data and for key opinion leaders (KOL) and testimonials of various sorts. Usability engineering, too, may offer some of the activities and documents needed as part of this evidence building, but more often we will be talking about the clinical evaluation report (CER) and its precursor the clinical evaluation plan (CEP). Clinical validation and validation of the usability of the device (i.e., summative evaluation) are, after all, not the same thing as pointed out in Section 1.15.

The ISO 14155 standard does a great job of levelling the playing field here so that the clinical environment and the manufacturer both understand what is expected of them. This also describes the role of ethical committees, investigators, study participants, sponsors, and the competent authorities overseeing it all. The document does not tell you everything you need to know – you should always look at local rules and regulations, for example – but everything it does say you should hear. Rather than attempting to discuss the standard at length here we will point you to it as further reading should this important pocket of activity interest you.

5.4 Co-developing new products

The two worlds, business and clinical medicine, may be at odds with each other: business folks stereotypically care about money, just as clinical professionals virtuously care about curing patients. At the same time, the two make for a great team where both a need and a solution to it meet.

If a product does not demonstrate utility or a benefit in the clinical setting, it will be left behind. Sometime manufacturers dream of their products reaching a “must have” status whereby going without the product is a risk scarier than paying for it even when it actually isn’t necessary. Some laboratory tests might fall into this category, or a device the patient knows to loudly demand. At the same time, if the bucks aren’t there or if the financial calculations don’t convince, a manufacturer will have a hard time committing to building something even if it would be easy for them, academically cool and even have value to a small group of patients. Thus, money and medicine will be intwined – and both aspects have to be satisfied for the conversation to continue.

At the beginning of the conversation, a concept may be what we have in mind. At this stage, industry folks will be looking to find problems that their solutions can solve as is or with as little modification as possible. The cliché we often use here is that a company wants to solve everything using a ready-made database engine they have in their back pocket, basically like a spreadsheet. This may be just what is needed, finding a tried-and-tested solution to a gnawing issue or even an unexpected improvement to a problem no one had even realized was an issue before. The result may, however, also be a prime example of forcing a round peg through a square hole just because you need to, and you can charge the same profits for this duct-taped solution than a real tailor-made one developed via greater resource expenditure. At the same time, the folks on the other side of the table may be asking for more and more pegs to be squeezed through that hole because they think they can get more for the same money by doing it all at once. In terms of usability the result may be horrible, it may be an exercise in ticking all the defined boxes but not really making anyone happy in the process. After some hackling a point will be reached after which you will no longer get more for the same price, only something nominally more with hidden cutbacks.

The above also applies to most consumer electronics out there: if a manufacturer makes a top-of-the-line product and a somewhat cheaper model you can bet the differences in features, build, and support over time are far greater than the few bullet points on a brochure at first suggest. The difference will become apparent sooner or later, but to who, when, and how painfully, are the real questions to ask. Occasionally, the top-of-the-line and the best-selling standard model may, of course, also be identical in every respect. With custom-made or tailored software that is not likely to be the case. As a side job, we used to sell software development projects and took pleasure in explaining to potential customers that we could build the software for X dollars or 2X dollars, but the result would not be the same software in both cases even if the features looked the same on paper. This often took people aback, but after some discussion they almost always appreciated the realization. With most products there is the “80% solution” developed with as little as 20% of the effort, and the 97% solution that took 100% effort to make. Often times something built quickly will be a pain to maintain over time as new features are requested or the libraries and components used begin to show their age. At the same time, perfection is an expensive illusion that might see the product permanently stuck in development.

Some companies may be looking to identify new issues to solve. Particularly startup-minded individuals might be scanning the field for new unmet needs. Developing a medical device, however, takes considerable time and resources so there should be a good match between the company and the need if we are to see results quickly or at all. In business, no one has infinite budgets or infinite deadlines. If something doesn't show promise in 3–6 months the clouds will begin to gather, and if in 6–12 months we are still twiddling our thumbs, or unsure how we will recoup our expenses and make a profit, the axe will probably already be taken out for sharpening. Transpose this against how a typical innovation cycle might look like in medicine: the physical principles behind magnetic resonance imaging were discovered in the 1940s, a concept for making use of these for medical imaging was proposed in the 1960s, and finally mature devices began appearing in the 1970s and 1980s – and only then did we start to build clinical data for the use of the technology. The lesson here is that for some, perhaps even most, new medical technologies it may take decades for the devices to mature and the required clinical evidence to be gathered. Different companies have different timeframes, different overall resources, and different interests in any given topic, but a wait and see approach probably means nothing will happen now and nothing is likely to happen later either. Time is of the essence is a cliché, but it happens to be true. In academia, on the other hand, it might be good enough to keep returning to some topic year after year like a pet topic. In the clinical context, based on our experience, the cliché is that we either do something or don't do it. In reality situations vary, but clichés exist for good reasons.

It's perhaps a fallacy to think that the manufacturer will do all the work and the clinical experts will just need to give a thumbs up to the right results. Some collaborations may work like this, but if the organization really wants to adopt the developed product, they too will need to invest effort into the long-term development, staff training, trials, patient enrollment, onboarding, use, and even the long-term maintenance of the product. This will particularly be the case if the product is something as complex as software based on artificial intelligence. Our point here is that not only is there no free lunch, but also out of all of the 20-dollar lunches you could have, only a few truly are win-win propositions for both their maker and consumer. See Juuso (2022) for more on outsourcing your development work, in that case for QMS development, but it all applies here too.

Companies don't often have the luxury of taking a novel concept and turning it into a finished product just to see if it could work. That no doubt happens, but if you have an idea for a medical device, you should also have an idea for its market to really get the attention of prospective manufacturers. You should also have an argument for why your new product would fit well into your clinical environment and the portfolio of devices that a particular manufacturer already makes. If you are the manufacturer with an idea, you should try to understand the healthcare environment you want to fit into as closely as possible.

In the end all that matters is that we somehow moved the needle and made the world a better place. In Juuso & Pöyhönen (2024) we briefly discussed the inspirational inception story of Medtronic which we feel serves as perhaps the foremost example of collaboration across the divide between manufacturers and clinicians. As discussed in the book, it was in 1957 when a University of Minnesota heart surgeon, Dr. C. Walton Lillehei, and an enterprising electrical engineer, Earl Bakken, dreamt up a wearable transistorized pacemaker and only days after attached a prototype to the first young patient. A year later the device was on the market, and decades later an implantable version of this same pacemaker is now commonplace. The world of medical devices is perhaps full of similar stories of real-world needs and the required skills being combined to bring new, much-needed solutions to the light of day. Yet, as a story of the manufacturing and clinical worlds colliding together for the betterment of care we can think of few stories as inspiring. The next story we read about, though, might be yours.

5.5 Bottom line

A medical device is always the goal of the product developers, but for clinical specialists the goal is treating a patient. The developers may fall in love with the technology inside the device whereas the clinician may not even want to understand all the particulars of how it works just as long as it works. The clinician's goal is a healthier patient or a vanquished disease, the mechanics of which may be their obsession just like technology is for the engineers, or they may just care about the result. In the clinical setting, a device that "just works" is much better than a device that calls attention to itself. Debugging a device is not something a clinician will be willing to do, and it would usually be time away from treating a patient.

It's also a fallacy to think that all medical professionals are alike. In reality, the depth, breadth, and focus of medical knowledge at the command of each doctor, nurse, or hospital physicist will vary greatly. A specialist doctor, a general practitioner, a substitute locum doctor coming in via a medical staffing agency, and a senior hospital physicist specialized in a particular advanced technology will all have different sweet spots of knowledge, experience, and interest. Occasionally language skills, too, will differ so that a new nurse or a doctor may only have a passing command of the primary language in the country they find themselves in, or conversely that list of languages at their command might not include the English we often take for granted. Cultural differences may be even more pronounced between different countries and organizations than just language alone: the administration structures, care flows, infrastructure, staff composition, and customs will be more diverse than what we have seen in the TV series "ER" centered around one fictional Chicago emergency room. Each user will have different characteristics for using a device safely – it is up to us as manufacturers to figure out if and how the devices can be handed over to these users.

The one universally appropriate simplification of this environment is that everyone is busy doing something else, or just getting things done, at the same time that we wished they would focus on understanding and using our device. Instructions for Use are probably good as a last line of defense and a curated reference of information on the device, but they are no guarantee of safe use by themselves. We have yet to come across a doctor who sleeps with a copy of such instructions by their bed, but we have seen stacks of manuals shelved in their original cellophane wrappers at a major hospital and we have heard doctors say that devices should just work out of the box on more instances than one.

After all the time we have spent traversing the divide between academia, industry, and the medical world we firmly believe that where there's a will there's a way. For a joint will to emerge there has to be a sense that the proposition makes sense to all involved, is doable, and is doable by the group of people assembled together. If you think you might have that, pickup the phone or shoot off that email and get boots on the ground.

Chapter 6

Benefit Management

Benefits have long been a factor in risk management where we have been instructed to weigh risks against benefits in some loosely standardized way as Figure 6.1 *attempts to illustrate. The right balance has largely relied on expert assessment while structured, classifiable, and quantifiable approaches have been the topic of much discussion but little standardization. Benefits are primarily thought to come from the clinical sphere of activities, but also usability may play at least a catalytic role in their realization.*

Benefits may occasionally feature as a sort of an afterthought to risk management activities or even the rubber stamp used to waive past whatever small residual risks are left after the adopted risk mitigation measures. This is a bit like the hero always surviving at the end of a movie so that a sequel may eventually be made. Ripley always battled her way through whatever horde of quadrupeds the alien queen sent her way. John McClane always found a crack to squeeze through before the final big explosion rocked the scene. Rambo always stood tall over the mayhem he had caused. James Bond, too, usually found a way to retreat from the scene with a beautiful co-survivor before the end credits rolled – and those credits then promised he would return with a new adventure. There is a definite benefit to the protagonist surviving, but as a benefit it's sufficiently utilitarian to be taken for granted. In medical devices, too, such a last-minute checkpoint may act to ensure we are still doing more good than harm with our device, but as an approach to risk reviews it may not be enough or even entirely appropriate to making good medical devices. Instead, we should genuinely understand the benefits associated with our

DOI: 10.4324/9781003645702-6

Figure 6.1 Benefits ought to be the driving factor for our medical device, not just an afterthought in risk management.

device – and understand these enough to both identify the appropriate benefits and to classify them in some appropriate way.

No device is built purely to minimize risks while paying no attention to any benefits sought. Otherwise, the best solution would be to not build the device at all and thus reduce all risks related to the existence of the device to zero. All devices should come with some benefits, and we as the developers, manufacturers, assessors, and even users of the devices should attempt to understand the benefits as well as the risks that come with each device. Understanding risks is often hard enough so that analyzing benefits may seem like too much to ask. If we, however, want to claim that some risk is worth taking or some risk can't be reduced further we will need to take a rigorous approach to benefits as well. In other words, it may not be possible to reduce a risk any further, but if there is no benefit to taking that risk it should probably be eliminated altogether.

The topic of benefit management is currently an area of noticeable activity in international medical-device standardization. No single model for carving out a bigger, better-defined role for benefits as part of risk management yet exists. Mechanisms for treating risks and benefits as numbers and simply calculating a ratio of these for the benefit/risk assessment have been suggested, but so far no consensus has been reached on their merits. Work is, however, ongoing to prompt a more structured approach to benefits than is currently commonplace in the field. Just like it is unacceptable to not think about the acceptability of risks overall, it's not going to be looked upon kindly if you use undetermined, unquantified, and unclassified benefits as a rubber stamp to brush away any risks you might have. More rigor is requested today, and much more rigor will be required sooner or later.

In this section we focus on usability-related risks, but the conversation is just as applicable to all other types of risks. The IEC 62366-1 standard on usability engineering does not delve into the topic of benefits, not exactly, but benefits are always there behind each sentence and word of the standard. The discussion in the standard is formulated around taking a piece of sandpaper and smoothing out any prickly opportunities for harm that our evolving devices may have so that they can be better put to their intended use and to reap the intended benefits. Yet, the standard routinely discusses, for example, users, use scenarios, and use environments. None of these have any real meaning without some utility, or benefit, associated with the use of the device. The language of the standard will evolve in the future, but the fundamental needs are already there even if in a few hundred years we might consider their wording antiquated – and even if the introduction of artificial intelligence promises to shoot us up into that future much sooner than a century or two. The point we want to make here is that benefits are not some strange new attractions perhaps coming soon to risk management, or that benefit management was some next-generation alternative to risk management, but that benefits and risks are both already here, they are always connected like the flip sides of a coin, and they must be addressed appropriately.

6.1 What is a benefit?

A risk is perhaps intuitive to understand. We have been subjected to the risk-based approach in the context of medical-device quality management long enough that the consideration of risks feels rather routine. Benefits, however, are a newer side of the same deliberations, having caught up with us gradually over the past years and become increasingly assertive in getting a voice in our discussion of risk management.

Benefits, too, can be intuitively addressed, but not all benefits we can think of should qualify for consideration in our assessment of the benefit-risk ratio. The first critical factor here is that the beneficiary of the benefit is the same as when we address risks: the user, the patient, or their environment somehow as expected for risks by ISO 14971. Let's call this the first rule of benefit management. Exulting the benefits reaped by the manufacturer while looking through our fingers at the risks facing users is exploitation. Plain and simple. Counting the number of rotten apples and comparing that to the number of edible oranges is not what the standards or the regulations ever want to see done. Where the application of this rule perhaps gets murkier is when we have multiple valid user groups, and the benefits might be somehow cross-linked. Even then the basic rule still applies, and we should not lose sight of the appropriate coupling of benefits and risks.

This brings us to a second clear-cut rule: business risks (e.g., financial risk to the manufacturer) are expelled from ISO 14971 risk management and so too should business benefits (e.g., the manufacturer can make the device cheaper). If, however, a cheaper cost means that the product is available to some new patient group that could otherwise not afford the device, this may be an appropriate benefit from the point of view of the patient. At the same time, it would be entirely inappropriate to cancel out real patient risks with financial benefits the manufacturer might reap – that is in fact what happened in the breast implant scandal where a manufacturer used industrial silicone in implants to save on the raw materials but ended up harming the health of the patients and inadvertently ushering in a new stricter regulation in the form of the EU 2017/745 medical device regulation ("MDR"). Also consider that changing the design of a product to cut costs may change the risks involved with the use of the device, but they may also change the usability of the device through changes in materials and components. Our message here is that if the design changes, for whatever reason, you should consider the effects of that change in terms of both risks and benefits.

The take-home message from both of the above rules is that the point of reference for any study of benefits and risks should be with the patient and the user. To do so is not a matter of wordsmithing the definitions in the risk matrix, but rather a genuine focus to adopt in analyzing risks and benefits.

The third rule will seem more technical in nature as here we want to ensure that the appropriate match is made between risks and benefits. From the evolving solutions we have seen over the years, individual risks are fairly often checked with some higher-level benefits. Call this putting out a smoking cigarette by dumping a cargo plane's worth of water on the unsuspecting employee on his cigarette break to prevent a forest fire. Instead, we would like to see both risks and benefits addressed on the same level of granularity and paired so that a particular risk in the matrix is intimately linked with a statement of benefit. This way it would be possible to evaluate the necessity of taking some risk in the first place, and then to review its mitigation. There is, of course, no single correct way of doing any of this which makes the ask tough all around.

Consider, for example, the case of a fictional automated external defibrillator (AED) intended to restart the human heart after cardiac arrest. Having such a device readily available nearby when an emergency occurs is self-evidently a benefit. It might also be a cool idea to have the device automatically call emergency services as soon as the shock pads are broken out of their case. This way an ambulance might be on its way minutes sooner. Yet, if a network connection could somehow deplete the battery on the device or if a network error or cyberattack could somehow prevent the appropriate shock from being administered, such a feature would not be seen as an improvement on the device. In such a case the network-related risks should not be countered by just having a device available for purchase or not. The appropriate benefit would have to be linked with the benefits the particular network-enabled features bring. Similar decisions will be needed if a feature can be realized in multiple different ways, say the use of a serrated blade versus a laser, or the use of artificial intelligence versus an experienced doctor to consult. Such consideration may be easy or hard, but they will need to be made to analyze risks and benefits on the same footing.

At a recent FDA-sponsored conference, a room full of risk and quality experts were asked to design risk management for a particular case study handed out to the 90-plus participants of the event. The information package was the same, even the risks described were the same for all participants, yet at the end of the day we had about as many unique classification matrices as there were groups of participants in the room. The lesson of the story is that finding two people who will analyze risks the same way will probably be as rare as finding two snailfish who both enjoy basking in the sun, but the point is to perform a systematic and considered review of the risks, benefits, and appropriate mitigations associated with the device we want to produce. Getting as many points of view from as many appropriate experts and stakeholders as is possible will probably improve the outcome, which is why many standards remember to ask for such diversity in participants to different reviews. Who the appropriate experts are will depend on the device in question.

Ultimately, not every risk can be overturned by claiming the device can't be made available without the individual risk. Some risks can but for others alternative designs or processes might lead to a better device or device still achieving the benefit sought. The point here is that each risk should be paired up with an appropriate benefit, not just the big rubber stamp we have at our disposal.

The fourth rule builds on the previous rule by asking that both risks and benefits are categorized, classified, or otherwise estimated using comparable scales. If we are to weigh benefits and risks with each other it doesn't help if one is stated in micrograms and the other using a fluid scale of how much we feel like addressing the topic. The classic scale used in risk management for risks is a 5x5 table set up by the probability and the severity of each harm. Probability may be appropriate to both risks and benefits, but the goal with benefits is probably to maximize rather than minimize the probability. Severity may perhaps be translated as criticality for the benefit sought, but no standard for this exists. Consider, for example, the following fictious classification of benefits and risks on a seemingly compatible 5-level scale given in Table 6.1.

In the table, risks are classified with the sort of labels commonly used in managing risks today. The same labels would have no meaning when applied to benefits – after all what would be a "catastrophic benefit"? An ecstatically fantastic, out-of-this-world benefit would also be a poor counterpart to a catastrophic benefit as it would be impossible to agree on what this means. Furthermore, we could quite easily assign marketing speak to all the benefit classifications and arrive at a situation where all the benefits have some positive, nice-to-have flavor to them. This might not be useful in a situation where perhaps hard choices have to be made on weighing benefits and harms. Just because it takes effort, though, is not reason enough to not tackle benefit management.

The fifth rule of benefit management we would introduce here is that, unlike risks, benefits should be realized over the use of a device. This turns post-market surveillance on its head if compared to a risk-based approach: instead of hoping we don't encounter realized risks over time we should do our best to tease out objective evidence on the realization of the benefits we envision and whatever else benefits are being enjoyed by

Table 6.1 Classification of benefits and risks

Benefit levels		*Risk severity*	
5	Out-of-this-world™ improvement	5	Catastrophic
4	Must-have improvement	4	Critical
3	Life-changing improvement	3	Serious
2	Game-changing improvement	2	Minor
1	Nice-to-have improvement	1	Negligible

our users. It will be infinitely harder to claim that "All the expected benefits were seen to occur in the wild" than it has presently been to somewhat off-handedly claim that "No new risks were seen in the wild". The good news here is that IEC 62366-1 usability engineering does provide tools to obtain evidence on exactly these topics (see Section 3). Similarly, the clinical benefits our clinical evaluation report (CER) spoke of should also be evident over time. This may be a topic for both post-market surveillance (PMS) and for post-market clinical follow-up (PMCF), but as discussed in Section 1.15, clinical validation and usability evaluation are not synonymous with each other. Thus, PMCF is perhaps better suited to maintaining visibility on the clinical benefits over time than usability engineering itself.

In the interest of benefits actually being realized over time, no mitigation on benefits is of course expected, as you would expect in the case of risks. It may, however, be practical to consider how the likelihood of some benefit being realized over time might be improved. This may impact design choices in shaping the product and its associated documentation, for example. This is product improvement, and it is something ISO 13485, for example, does expect to take place.

Thus, the top five rules we suggest you consider when addressing benefit management are as follows:

1) The beneficiary of the benefit and the object of harm are the same
2) Business benefits are not eligible as such
3) Benefits and risks are addressed on the same level of granularity
4) Benefits and risks are classified on comparable scales
5) Benefits should be realized in a verifiable manner

Benefit management is an evolving topic that will see many more helpful and no-doubt further-developed frameworks emerge over time than the one we presented here. The five simple rules we gave are, however, important aspects of addressing benefits that we feel are not always readily observed in discussions today.

6.2 Approaches to managing benefits

The current approach to benefits as instructed by ISO 14971 risk management is to focus on risks, but then also consider benefits somewhere between the lines and especially at the end of the day. In a way this is entirely appropriate: we focus on risks as those are the sharks in the water that could get in the way of our users realizing the benefits we otherwise envisage. We don't focus on the dolphins joyfully galivanting around the oceans or the snailfish habiting the bottom of the sea. At the same time, however, this is like treating Jaws as a film about fishing and neglecting all the characters, sunbathing, and general merriment that occupy the time between the few shark attacks Spielberg's flailing mechanical beast can muster. The benefit that was sought with the film was entertainment and excitement for the audience, for whom the risks probably included little more than perhaps being momentarily afraid of dipping in the sea or the family bathtub. The benefits probably outweighed the risk of aquaphobia quite clearly for most viewers.

ISO 14971 asks that we, when devising risk mitigations (i.e. risk controls), also consider the benefits we had in mind. It fears that while going through the motions of managing risks we neglect to check whether the benefit is still there after all the protective padding and warning stickers we have lavished on our design. Staying onshore for fear of sharks will mean no swimming. Instructing everyone to stay home in their basements will also mean no sunbathing. The risks and benefits ought to be somehow measured in each case, and it is up to you to ensure that they are – and that you can then make this case to others when they come asking.

Going a step deeper, after all risk mitigations are performed and we are left with some overall residual risk we then get to review all the benefits of our device and weigh these against all the risks. This ratio of benefits to risks used to be called a risk-benefit ratio, but after the 2019 revision of the risk management standard this was flipped to benefit-risk ratio in order to put benefits front and center. The writers of the standard thus want us to consider the benefits sought and then assess whether those still outweigh the risks we fear. If the reason we wanted to dive into a shark-infested ocean was just to cool down on a hot summer day the prospect may now appear somewhat reduced in appeal. If, on the other hand, we took a calculated risk to attract ten million subscribers on YouTube, and thus perhaps earn a buck or two, maybe the prospect of losing a limb or two appears marginally more appealing.

No method or classification is in fact given by any of the standards for the assessment of benefits. No simple formula exists for how much bigger the benefits ought to be than the risks for us to feel happy about moving ahead with a product launch. There is an expectation that our clinical evaluation, particularly the all-important clinical evaluation report (CER), will provide evidence for the benefits we invoke in risk management. Simultaneously, sitting patiently in the wings of our stage is the IEC 62366-1 usability engineering standard biting its tongue on the fact that also usability might offer some real benefits to the users of medical devices – not just as a way of thwarting off harm but as a way to provide genuine benefits in the first place. An example here might be, for example, increased adherence by patients to using the device prescribed to them as a much safer and better user experience is achieved via usability engineering. None of this is voiced too audibly by either standard, or by any other standard, and thus the manufacturer is largely left to ponder benefits and argue their relation to risks in their own way and as befits each device.

Occasionally, discussion is struck up in standardization circles about how to best classify and analyze risks in relation to benefits, but this has not yet led to any breakthroughs. A French model has, however, been presented in the past few years and may lead to better tools in addressing benefits as part of ISO 14971 risk management. This work seems set to influence a future revision of the optional ISO 24971-1 guidance document on risk management. Regardless of this revision work, no mathematical ratio or a single number for the benefit-risk ratio is expected by the standards. Occasionally we hear that such a figure is insisted on by some junior auditors, but there is no such requirement in the standard that we have been able to identify. If used inappropriately, such a figure might actually do more harm than good, as explained in the following section.

6.3 Ultimate review

The world of standards and regulations is written from the point of view that after weighing all the evidence, the manufacturer comes to a final conclusion on the benefits and risks involved in the use of the medical device, and either decides to proceed to launch or go back to the drawing board somehow. The manufacturer has full access to all the data, knows what kind of brushes and carpets seemed appealing at what stage, and what was the edge of the surveyed universe in risk assessment. The intention here is that after having seen everything, measured everything, listened to all the experts and the users, and having weighed it all, the manufacturer makes an assessment and is confident in defending that decision including by providing appropriate objective evidence.

The notion of creating a simple mathematical formula for weighing risks and benefits is an appealing one, but it is ultimately a flawed approach if it chips away at the review by the manufacturer. If the formula comes back with a result that benefits are 0.2 units greater than risks, and the manufacturer uses this result as the traffic light to go on, we run the risk of running tire marks across the standard and the regulations. As beautiful as that formula may seem it is a Trojan horse and a false god if it stops us in any way from taking the time to review what the benefits and risks actually are or if it lets us skirt responsibility in any way. The standard is big on responsibility, and so are the regulations. If you want to let ChatGPT make the final decision on publishing your device, you are undoubtedly cutting-edge, but you are also uncomfortably close to the chopping block with your device.

It is one thing to have perhaps questionable quantitative data, and trying to use that in your risk assessment, but it is quite another to not even look at that data, or believe that a formula used by someone else would give anything more than a kneejerk calculation on the variables you happened to set. There is certainly a right way to use quantitative data and formulas, but the risk of using them incorrectly is quite high. If your data is perfect and if your assessment of individual risks is immaculate, perhaps a formula could be the right way to go as supporting evidence for your expert assessment. If your data is based on interviews and observations of 15 test users, or anecdotal media reports of what your competitors have done, we would be hesitant to call the data quantitative. Risk management in the whole is not yet an automated process, and perhaps it never should be that entirely.

For more on risk management please see our earlier books Juuso (2022) for risk management of medical devices and Juuso & Pöyhönen (2024) for risk management in the context of medical software.

Chapter 7

Practical Transparency

In movies, the Hollow Man is transparent to the point of invisibility, and although see-through to the extreme, perhaps not exactly a poster child of practicality or actionability in opening up to your fellow men. On the other hand, eye-exams are a tried and tested mechanism for either seeing or not seeing something, as illustrated by Figure 7.1, *but to know what level of transparency is called for, and what that information disclosure should involve, some other type of test is needed.*

The traditional goal of usability engineering is to understand the user. Most of the landmark literature on the topic focuses on how we should approach the user, observe their actions, model their interaction with our devices, and work to ensure we know how to best meet their needs. This is appropriate, proper, and beneficial for developing useful, safe devices. Over the years our understanding of the devices has gradually evolved from just the device itself to the information we expect to see accompany it: the user manuals, quick start guides, cheat sheets, and many other forms of documentation found "in the box". Observing, measuring, understanding, and improving the readability of this documentation and its usefulness has always been a part of usability engineering activities.

Yet, the situation has always fundamentally been one of what you see is what you get, 'WYSIWYG'. The device in front of the user has been open to interaction with the user and the material at hand has explained how it could be manipulated. Working a scalpel was something a surgeon needed no instruction on from its manufacturer, but printing out a patient discharge record may have been instructed through a few bullet points to the new intern starting work at a hospital ward. Software has gradually pushed the

DOI: 10.4324/9781003645702-7

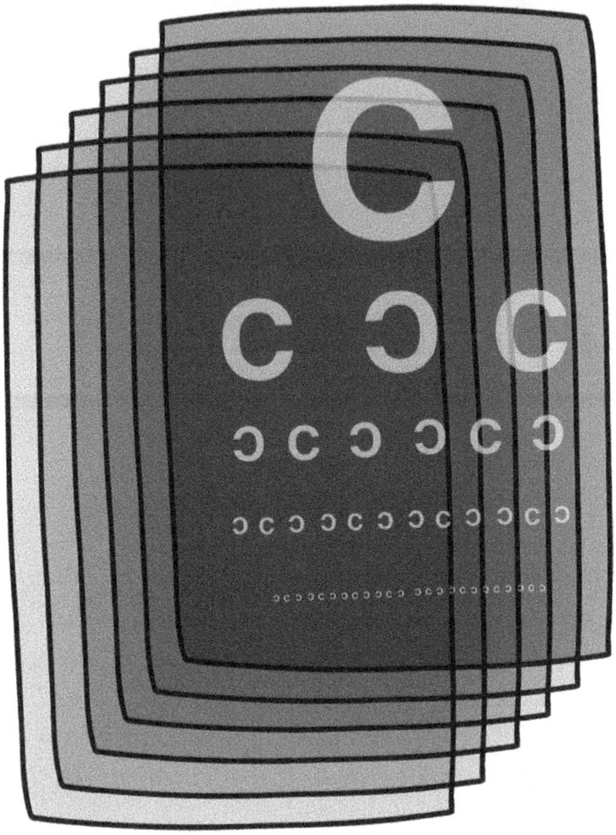

Figure 7.1 More than an eye-exam is needed when we want to verify the appropriateness of information.

envelope on what information should be provided and how this should be presented to the user, but the recent march forward of artificial intelligence technologies into medical devices has greatly expanded the need to explain the relevant inner working of the devices to their users.

AI-based medical devices may be white boxes where the mechanisms at play are perhaps well-understood and based on scientific knowledge, or they may be black boxes where we have a reasonable expectation that the utility we are seeing from the devices is based on appropriate characteristics of the underlying data, but we might not know this for certain. What is certain is that these new types of medical devices are much less 'WYSIWYG' than their predecessors while we expect much greater utility out of them than say from a scalpel. With great power comes great responsibility, though, as Spider-Man would say. In medical devices the need to provide useful, actionable information on the devices in a practical way has never been greater, and it is likely to only become more pronounced as devices evolve.

7.1 The go-to solution to transparency

Transparency is today an almost universally accepted goal for advanced software-based medical devices. Achieving transparency is, however, not quite as easy as with, for example, the electrified switchable glass that may turn the walls of a Tokyo toilet from opaque to fully transparent at the flick of a switch. The toilet, too, has suffered some publicized technical issues of late, but these pale in comparison to the headscratchers we are faced with in enabling transparency over medical devices. Let's next look at a few issues and possible solutions in this regard.

7.1.1 Getting all the information is the answer

It is all too easy to ask manufacturers to provide more data and more information on their devices. More of each is no doubt needed as we go forward, but quantity alone is not the solution. From the manufacturer's point of view, it will never be possible to provide all the information and data. The manufacturer has invested perhaps heavily in collecting and compiling the dataset and is naturally hesitant to hand it over to the public domain where all of their would-be competitors would be free to use the same data. This is not just an intellectual-property issue or an issue of trade secrets either, the calls for privacy and anonymization of the data are also important considerations here and sometimes run contrary to expectations of traceability and well-founded secondary use of data.

From the user's point of view, getting all the data would in most cases be a bigger burden than some help in understanding the device. Think of the usual terms of use we routinely see on websites and on splash screens when first firing up some new piece of software. How many of us actually read the terms line by line and digest the information given to us with scientific meticulousness? Hardly anyone and hardly ever, will be the honest answer here. Yet, that is the kiddie version of the pool we are proposing to throw medical-device users into. You can claim that we would be more inclined to read the small print on a medical device than on some online game, which is true enough, but still not likely to withstand the test of time and avalanche of updates and competing offerings. Shipping boxes and boxes full of documents to the opposing counsel may be a sound strategy in some legal thriller, but it is not a user-friendly tactic in medical devices. Resorting to the same tactic of inundation time and time again should earn you some contempt from your users and the court of public opinion at least.

Asking the medical professional or the patients to immerse themselves in all of the development data, documentation, and science behind some computationally advanced model is just a high-brow version of hiding from the issue of explaining how a device works and what it is good for. Some may make sense out of the material, but many others will just throw up their hands and either go with the flow or go away.

The worst part of providing all of the raw material is not that it might be so overwhelming that it would not be used by anyone, but that it might be incorrectly used by some. If the recipient of all that material was not in a position to decipher all the information correctly and reach the correct conclusions based on it, we might be courting misinterpretation and adding to the risk of misuse of the device beyond its original intended use. This might lead to misinformation, misuse, and unnecessary litigation when the goal was just the opposite. In an ideal world providing all the information could do no harm, but the world we live in is not ideal.

None of the above should be interpreted as a call to withhold information. Quite the opposite. We, however, need to find practical solutions to sharing actionable information on the devices so that the users can feel safe in their use of the devices and also accept their share of responsibility over that use. The manufacturer will always be responsible, but so too should the user.

The call to be given more data and more information is a fairly safe rallying call in that it will never expire. There will always be more data to be had, but the question should be what do we need to know and why don't we yet know it? This will help us identify the key pieces of the information puzzle, and work to both provide and obtain that information in as practical a way as possible.

7.1.2 Regulation is the answer

Similar to the call to be given all the information is a call for regulations to force manufacturers to give out certain information. Regulations will no doubt be needed to jolt all manufacturers to providing the same sort of information in some form resembling mutually comparable information, but we are probably not there yet.

At the moment the whole world appears to be interested in understanding new complex medical devices and especially the artificial intelligence aspects of these. Everyone appears to be in agreement that information exchange needs to take place, but we are only switching on the first searchlights for figuring out what that information needs to be. Mandating some information to be disseminated is a premature step when the information needed is shrouded in fog and the necessary background information, context of the data, and all interoperability requirements appear to be as distant and unknown as the western passage to India for the contemporaries of Christopher Columbus. Once we know what information we already have, what more we need, and how we can safeguard the appropriate interpretation of that information, we can mandate its disclosure.

The appropriate way to disclose this information, or the repositories needed for such disclosures to take place, also remain an issue to solve. In Europe the new EUDAMED database of medical devices offers itself as a self-evident solution, but it's still a long off from meeting its present regulatory needs, let alone taking on new ones. In the US, the FDA is very good about providing access to databases and datasets of interest to the users of medical devices. It too might balk at the sudden onslaught of noncurated, incompatible datasets it would have to host indefinitely and provide manufacturers access into for necessary updates. In between, there are several researcher-led catalogues of medical devices and even the digital health technology assessment databases run by various countries across the world. These are commendable initiatives at figuring out the real-world needs for describing and explaining medical devices, but forcing manufacturers to interface with all or several of these is not going to be feasible. There is also a natural question mark around how the properties and fields utilized by each of these initiatives map to each other, and how they should be interpreted when comparing entries across sources or when manufacturers feed data into multiple catalogs.

Regulations will be a part of the answer, but we should really know what to ask for before forcing the industry to provide it. We can, of course, demand the manufacturers disclose all relevant information for the safe use of their devices, we actually already essentially do this, but until we are able to formulate the request in more concrete terms, we won't have a great expectation of uniformity, compatibility, and comparability of the information from one manufacturer to the next.

7.1.3 Standards are the answer

You saw this coming, right? We have written a few books now on standards, so the topic was bound to come up as a solution here. The fact is that standards have gone a long way into standardizing the way manufacturers operate and develop medical devices. Knowing that a manufacturer adheres to ISO 13485 quality management will give us great insight into how they address various stakeholders in the domain, how they approach product development and the long-term support of their customers in using those products. Furthermore, knowing that the manufacturer is expected to comply with, for example, ISO 14971 risk management will give us a reasonable expectation of how they approach the identification and control of risks possibly related to the use of their devices.

Most users are oblivious to these standards, but they don't have to be intimately knowledgeable of their content to appreciate what the standards say about reasonable expectations between them and the makers of the devices they interact with. Product standards directly speak to the properties of the devices and allow for some common bar to be raised across the different vendors of a particular type of device. For example, claimed compliance to the IEC 60601-1 family of standards gives us reasonable expectations on the electrical safety of a device just as conformity to IEC 62304 raises an expectation of professional software development having gone into the manufacture of a product. Organization-level standards such as ISO 13485 then augment this framework to give us the handles, or the application programming interface (API), to understand the mechanisms of the organization behind the device. All these standards will be key to predicting actions and relying on their output as the products put out by the organizations evolve over time.

We also already have standards and technical reports on the information we expect manufacturers to provide on their devices. ISO 20417 on information to be supplied by the manufacturer is just such a standard. ISO 15223-1 goes a step further and gives us the graphical symbols to be used with such information so that a user can, regardless of their language, grasp key information on the device, its manufacturer, and, for example, the expiration date of the device in their hand.

We also have further standards on how the manufacturer is to keep an eye on their device after it leaves their factory floor. The ISO 13485, ISO 14971 and IEC 62304 standards mentioned above include such expectations, but ISO TR 20416 is a dedicated document on the expectations around post-market surveillance performed by the manufacturer. In all, there are some 24,000 ISO standards in existence today. No more than a handful of those are likely to apply to a typical medical-device manufacturer, but those that do offer, anyone who cares to take note, some real insight on the activities and products of the manufacturers.

7.1.4 ChatAI is the answer

24,000 ISO standards and boxes and boxes of documents, you say? That sounds like good fodder for the likes of ChatGPT, no? Perhaps. It may be possible to feed all the information to a generative AI system and ask it all the questions we want instead of going through the pain of a deep dive into the raw data ourselves. Perhaps someday AI will be able to give us reliable answers, but for now the rogue hallucinations and unwavering self-certainty of such AI-based conversationalists does not instill confidence. We would not yet ask AI to tell us if a mushroom in the forest is safe to eat or if it might kill us then and there. The image-based machine-learning algorithms available for such applications are certainly already very convincing, but the benefit and risk do not necessarily align favorably here. Similarly, we would not ask AI to recommend the best herbal cure for cancer and cancel an appointment with a true oncologist. Trusting the AI-generated top five easy steps to configuring my own radiation dosage during radiotherapy also sounds like a Darwin-award moment perhaps given out to the self-extinction action of the year.

Asking a generative AI system to munch its way through all the material and arrive at the whole truth and nothing but the truth behind it, is for now a distant application. It is, however, an application we are likely to see in our lifetimes. As a tool of convenience, it is compelling, but as a mechanism for taking responsibility into our own hands it appears backward minded. AI is not God, and blind trust here is the very opposite of understanding. As a solution to fostering transparency, generative AI may be a part of the answer and help us dig into that overwhelming stack of document boxes that arrived with our medical device, but it is not yet the prophet to blindly lead us out of the document era.

7.1.5 Education is the answer

Once you know, you know, right? Ensuring appropriate information and training is available to the user, and that any hard prerequisites on education and skills for the users of the medical device are defined appropriately are absolutely going to be a part of facilitating practical transparency.

The expectation is that instructions for use are written in a way appropriate to the users, and this should not just be understood as meaning the language is English or whatever language happens to mostly be spoken in the intended use environment. The user's context may, after all, be defined more by the local standard of care, any local hospital care practices, and the fitness of such instructions into that context than by language alone. IEC 62366-1 usability evaluation of such information is not just about language; it is about a sufficiently lossless understanding between the manufacturer and the users of their device. In the case of AI, the topics in this understanding are more complicated and nuanced than might be the case with some more traditional technology. Education is perhaps needed on both sides of the aisle, and especially on the topics that should be discussed for such an understanding to emerge. Asking all users to become proficient in AI is not going to be feasible. Asking all manufacturers to oversee their users like school principals is not going to work either. In practice, the responsible organization (see Section 1.8) will need to step up here, but not to shoulder the sole responsibility.

In the end, the real question is not whether education will be a part of the solution, but instead how the implementation of education will shape out in the real world. The manufacturer, the responsible organization, professional users (including doctors and nurses), and patients will form a hierarchy of knowledge and needs, and somehow we will need to make the appropriate information propagate through this network. The manufacturer is responsible for feeding the right information and device updates into this ecosystem, but in the end the patient (or their insurance or society at large) will pay for it all. The professional users have direct contact with the patient but asking all doctors to become AI experts or to use a significant part of the very limited time they have to spend with patients to educate patients on AI may not be the wisest of choices either. In this environment it may therefore be the responsible organization who needs to step up to provide information resources on at least some of the AI applications they run.

7.1.6 Simplicity is the answer

Simplicity is, of course, always a part of the solution, and at the same time it isn't enough of an answer. Remember the example of the on/off toggle for the respirator given in Section 1.9? "Keep It Simple, Stupid!" may be at the heart of the often-quoted KISS-principle, but at other times the solution adopted may just be stupid, as the respirator example proves.

A god-like AI system providing clear, concise, and to-the-point answers to our complex questions may feel like a great goal to have. Much like the oracles of the Greek antiquity, such an AI system could tell us with absolute certainty how to treat some ailment troubling us – or it could just as well just respond by a hallucination or by quoting "42" (see Juuso & Pöyhönen 2024). Unless we make the effort to understand the 'why' behind such answers, we will not be able to enjoy the full benefit of the answer and over time we may even end up decreasing our ability to understand, reap benefits, and even evolve. This is perhaps akin to a major corporation first outsourcing a part of its activities, then outsourcing the expertise to oversee and purchase those activities, and finally realizing that it no longer has any real expertise in that specialist field, which is actually quite important to its overall activities.

To some extent, applying AI-based devices should be as easy as using a word processor. In delivering healthcare services, the users should not be expected to be AI gurus, but those making advanced use of such technologies should not be treated like dummies and should learn more about the technology over time. In other words, the user interfaces we give to such AI-infused devices should not be overly complicated, but they should not be built as dumbed-down one-arm-bandits either. Designing such devices to be somewhere between an Enigma encoder and the Bombe computer used to crack the Enigma code may be a good goal here. Something used with sufficient training instead of the wizardry of Alan Turing at least sounds like an accomplishable goal.

7.2 The hierarchy of information

Enabling practical transparency for AI-based medical devices is a great topic perhaps worth a book of its own, but here we can't resist briefly tackling the status quo. Today, medical devices of virtually any type, including those infused with AI-capabilities, are expected to be developed according to a controlled process and pass conformity assessments before entering the market. The devices are expected to be accompanied by a set of documentation relevant to the type of device (see Section 4.4). Advertisements and scientific publications may then add to our understanding of what the device is and does. Let's look at this layered cake of information in a bit more detail next, from the outside in.

On the outer layer of the device's gravitational pull, some of the information may come from the advertising for the device, or from white papers published by the manufacturer and scientific articles published by them or others. Also, similar information available for competing products may fuse into the expectations the user has on a particular device. This information is, in theory, available to all, but not everyone will look for it or perhaps understand the material in the same way. Manufacturers will, naturally, hope that such material acts as the very point of introduction to the device for new customers, but it may also be a resource that interested users and concerned regulators dive into.

Diving deeper towards the core of the device, a curated collection of information on the device will be made available by the manufacturer to the users of the device as so-called accompanying information. This primarily consists of instructions for use, quick guides, release notes and the like, but will also contain the label and declaration of conformity regulators expect users to be interested in. This information is analyzed during conformity assessment and some of it will be very useful to the users, whether those users are laypersons or professionals. The information is public, but it's distributed to the users of the device along with the device, and thus not generally available to the public at large.

The next-inner layer of documentation consists of confidential documentation internal to the manufacturer. This will comprise both quality-management and product-development documentation (see Juuso 2022, Juuso & Pöyhönen 2024) that is generated and used by the manufacturer during the development of a device. The information is scrutinized during conformity assessments, but almost without exception not publicly available – although the Freedom of Information Act in the US has made snippets of technical documents available to the public. This documentation should, in theory, be able to answer every possible question on a device, but it is verbose, confidential, and perhaps even indecipherable to uninitiated readers not familiar with the standards and the manufacturer's processes. This is the bedrock on which conformity assessments are largely founded on, but for most other users this information is not available nor is it immediately clear that they would benefit from having access.

The innermost layer of information reflects the outcomes of conformity assessment activities and other regulatory actions. Various types of assessments, reviews, and reports are what we mostly have in mind here, but also the certificates written at the end of assessments or the official letters and notices written as a result of some event surrounding a device or its manufacturer will fit in here. Most of the detailed information will be confidential, locked away into regulatory databases, but, for example, certificates, letters, and notices may be made public.

Table 7.1 lists the layers of information in order from the inside out. The table also attempts to characterize the relationship of each key stakeholder group to the type of information, although in reality such relationships are a little more complicated than the simplified picture painted here.

7.3 Pushing the envelope on transparency

The previous section provided an overview of how information is structured today, and what sort of information exists at each layer that could be revealed in an attempt to foster transparency. Making all the information public will not be practical, possible, or in most cases even beneficial to any user. Despite this, we see that just like quality management and product development have increasingly led to databases of information and almost continuously moving processes, much of that activity may now increasingly extend over to the side of the user. Much of today's customer-related activities will inevitably benefit from and even rely on information generated deep in the bowels of the manufacturer.

No one correct model for bubbling up information exists, and the expectations around this are vague and loosely defined, but we expect the needs here to grow as devices become increasingly evolutionary and perhaps harder to grasp on intuition and quick guides alone. Artificial intelligence is just the opening salvo on this front.

Taking the red pill here and diving deep into the rabbit hole after Alice, we could dream up a sci-fi future where an ultimate level of practical transparency is achieved for a device. Riffing off the Jim Carrey film "Liar Liar", where a lawyer is unable to lie for 24 hours, we could envision a future device noticing that it is not suited to the use its user is attempting to put it to, and somehow speaking out on this. This is in fact already done

Table 7.1 Types of information and their use by stakeholders

	Document	*Layperson as user*	*Professional user*	*Regulator & NB*
Compliance	**Conformity assessments**	**No access.** May look for signs of CE-marking or device clearance.	**No access.** May look for signs of certification (e.g., ISO 13485, ISO 27001), CE marking, and device registrations.	**Source of information.** Most material is strictly confidential, but some parts may also be public (incl. US FOIA).
Internal documentation	**QMS documentation** (incl. CAPA, management review, audits)	**No access.**	**No access.** May look elsewhere for certain related certifications (e.g., ISO 13485, ISO 27001).	**Full access on demand.** In conformity assessment used both for initial status and changes over time.
	Development documentation (incl. D&D reviews, change management, problem resolution)	**No access.** Typically not a factor.	**No access.** May look for results of studies or usability evaluations.	**Full access on demand.** In conformity assessment used both for initial status and changes over time.
Accompanying information	**Instructions for use**	**Full access.** Information should be written for this audience.	**Full access.** Information should be written for this audience.	**Full access.** Compare to intended use, marketing.
	Release notes	**Full access.** Unlikely to look for changes unless critical use.	**Full access.** Should look for changes. Information perhaps written for this audience.	**Full access.** Will scrutinize especially along with changes. Information perhaps written for this audience.

	Residual risks and anomalies	**Full access.** Unlikely to read unless critical use.	**Full access.** May look for issues if use critical	**Full access.** Will scrutinize. Information perhaps written for this audience.
	Quick guides	**Full access.** May look at.	**Full access.** May look at.	**Full access.** May analyze.
	Declaration of conformity	**Full access.** Probably won't read.	**Full access.** Place for listing certifications.	**Full access.** Will scrutinize. Information perhaps written for this audience.
	Label	**Full access.** May use if use is critical.	**Full access.** Will look at the information.	**Full access.** Will scrutinize. Information perhaps written for this audience.
Public material	Advertising	**Indirect access.** May look at. May also be a point of entry.	**Indirect access.** May look at. May also be a point of entry.	**Indirect access.** Will try to look at, may request access.
	Whitepapers	**Indirect access.** Unlikely to read.	**Indirect access.** May look at. May also be a point of entry.	**Indirect access.** Will try to look at, may request access.
	Scientific publications	**Indirect access.** Unlikely to read.	**Indirect access.** May look at. May also be a point of entry.	**Direct or indirect access.** Claimed clinical evidence will be studied carefully, otherwise may request access.

via, for example, preprogrammed safety and input data checks on devices today, and while the devices usually only speak out via dialogues and prompts, the utterances may be quite important for safety and also fairly revealing on the inner workings of the device.

Fusing together the status quo of internet of things (IoT) -technology, present-day FDA-databases, and what the Star Trek films have shown us of the hive mind of the Borg, we could prophesize that in the future all adverse events, use errors, and close calls flow into a global database of incidents. Once trends in the data flows turn red, the devices could politely warn their users of the concerns or even refuse to work if that was the safer alternative. Today, the IMDRF is hard at work enabling such a future by fostering interoperability between regulators starting from adverse event classification in databases across the globe.

Borrowing a create or two from the Indiana Jones warehouse, we might pool together all the information on a particular device and let the user use it to their heart's content. Starting from the label, the user might use augmented reality to click through to the declaration of conformity, the assessment documentation behind it, the development documentation behind this, the quality management documentation underneath all this, and then branch out to any other related piece of information on processes, activities, and events following their trail of thought. We may have come a good way toward this utopian vision since the days of the first commercial medical devices, but we are not there yet. We might also caution that with this unprecedented access and power also come with responsibilities to ensure we as users understand the information we encounter and don't just create our own curated silos of false information.

Returning to a more blissful vision of a transparent future, we might paint a rosy picture where all the questions have already been asked and answered by AI. Such AI has gone through all the evidence, conjured up the last remaining missing pieces, and now confidently predicts all our needs of transparency into a medical device. In effect, this is more than a little reminiscent of the Jedi mind trick Obi-Wan Kenobi uses to pacify inquisitive guards in Star Wars by merely waving his hand and saying with a calm, steady voice something to the effect of there's nothing to see. From our chosen anecdote here, you can probably tell how we feel about relinquishing control over to AI too soon and too easily, but it is perhaps the logical conclusion to transparency: complete opaqueness, albeit with a benevolent benefactor pulling the strings. We could say that we are not there yet, but isn't this what we stereotypically look toward doctors for?

Someone to say that there is nothing wrong, or, if there is, that it will be fixed momentarily. To expect devices to do the same is perhaps only a matter of time, even if human doctors will hopefully not disappear any time soon.

We hope the above discussion gives you some ideas on how both information and information needs can map across the functions of a manufacturer, their users, and the regulators overseeing such use. We apologize if the heading of this section promised simple answers here when none are yet available. We are not there yet.

Chapter 8

Listening to the User

Seattle had Frasier Crane to listen to its worries, before him there was the Midnight Caller in San Francisco, and maybe even Sleepless in Seattle *would qualify as an example of making the world a better place in some small way through the act of tuning in. In medical devices we are forced to listen, and forced to have processes for listening as well as reacting. It's the first of these processes here that is perhaps too often prone to perfunctory lip service and archived records accumulating spider webs.*

At the beginning of our medical-device startup career we attended a lot of startup pitches, and gave a good few ourselves. In just about all the pitches the eager pitchers were trying to throw out absolutely everything out there in the hopes that something might stick. A five-minute pitch was jam-packed with far-fetched notions of business potential, incredible market size, winning team, game-changing technology, and whatever else anyone had ever thought to ask for during some previous pitch session. Everything but the kitchen sink, in other words. We always felt that more information equaled to more noise around the hook you wanted to sell. Too much noise and the best hook for that particular audience would never be picked up on. Neither of us professes to being any great pitcher on stage, but we always wanted to know our audience, understand their circumstances and what they were looking for in a pitch. Based on this, we could craft a pitch that left out 80% of the other stuff and put the spotlight firmly on the hook we wanted to cast. A 30-second hook to get us the next 5 minutes, which could lead to the next hour, and even a lifetime relationship. The 80% would work itself out over time anyway, we thought. The situations of

DOI: 10.4324/9781003645702-8

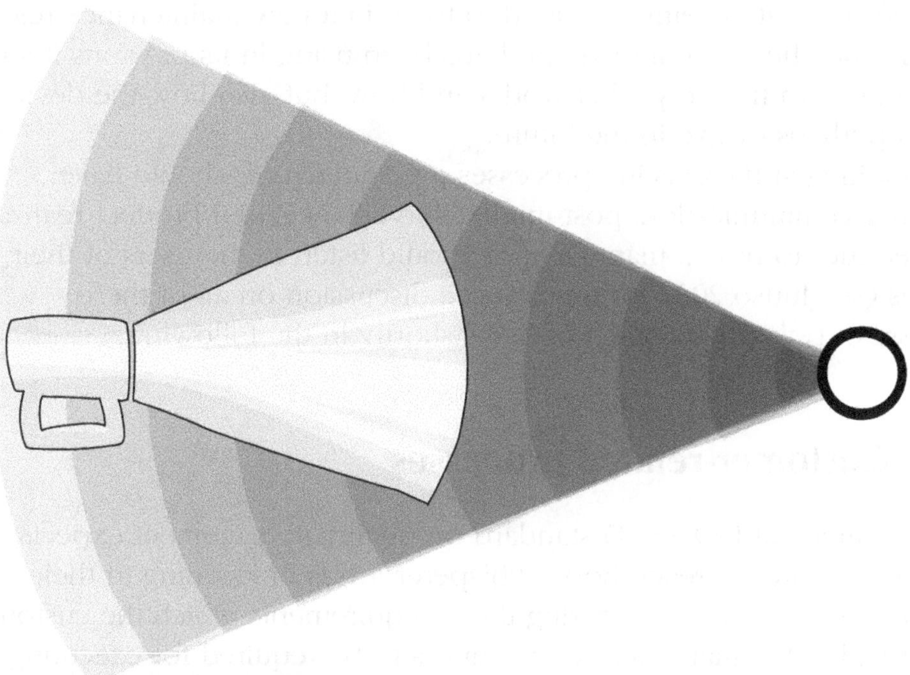

Figure 8.1 Receive and transmit, don't just broadcast.

course differ, and a long conversation is not always your objective, but a long-term win-win outcome always should be.

We feel that most people think selling is mostly about talking, talking the customer into submission and accepting the deal put in front of them. We don't profess to being any great salesmen either, but we always thought that selling, and selling a good win-win solution in particular, is at least as much about listening as it is about talking. That is where we also see usability engineering heading next. Usability is no longer just about pitching a device to the user and expecting that whatever explanation needs to take place for them to feel confident in their use of the device can take place in one sitting. The more complex our medical devices become, and the more the devices evolve through updates over time, the greater the need to enable some long-form dialogue between the user and the manufacturer – and the greater the need will be to not just broadcast information but to listen for the response. This takes post-market surveillance into the world of radar tracking, listening for the echo when releasing a new advanced feature for the customers to use. We don't mean to suggest trying out untested features on the users, not in the least, but somehow attempting to ensure that the users grasp what is being provided to them in a new maintenance release of a product they have already perhaps been using in its previous iteration. We might learn not only what works and how, but also how the devices should perhaps evolve in the future.

In looking at the existing processes a manufacturer should have, customer communication, post-market surveillance, and product realization all offer cues to how a manufacturer should listen to the users of their devices (see Juuso 2022 for much more discussion on all of these processes). Let's discuss each of these shortly in the following.

8.1 Customer-related processes

The fundamental ISO 13485 standard on quality management expects the manufacturer to reach horse-whisperer levels of listening to their prospective users by also hearing those requirements which the customers don't say, but the manufacturer can guess to be required for executing on the intended use the customers have in mind. The requirement is written from a general perspective of developing a device, selling it, and then having the device in use after that, but sooner or later the expectation will be explicitly applied to maintenance also. In fact, maintenance information as a form of feedback and design and development changes are already in the present 2016-version of the said standard. More advice is also perhaps forthcoming in the new associated ISO 23485 guidance document.

Examples of listening to users and potential users are organizing customer training events, attending industry forums and scientific conferences, meeting with relevant patient advocacy groups, and interviewing, polling or just talking with users. All of the above activities would qualify as active methods for listening to the user while all the classics discussed above – feedback handling and the use of maintenance information – might be more passive in nature. The use of logs and metrics might qualify as active methods, depending on how and when they are used.

8.2 Product realization

Product realization is an extension of the previous principle and culminates in the validation of the customer needs the manufacturer managed to extract. The design and development phase will involve requirements specification, design, implementation, testing, verification and validation which may all make use of access to potential users and other subject matter experts. The work will also feed into risk management, usability evaluation and the generation of clinical evidence, all of which rely on access to real users. These activities form a connected web of needs and solutions, all of which somehow affect or originate with the user.

Over the long term, this also includes the maintenance of devices, particularly in terms of providing updates to the products. How these updates are planned and executed should pay a similar level of attention to the user as the original product unless an extraordinary level of understanding of the user was reached during the original development phase. As alluded to above, looking at maintenance logs and other information gained from servicing a device will also be of interest with the mindset that, for example, seeing how a device has been broken will be telling of its use. Here maintenance information and in particular usage logs for software may be useful for both finding out if the device was used as we thought, or how we should perhaps change our views on its use. Facilitating access to such information must be based on consent from the user, but the access may be to the benefit of both parties. ISO 13485 expects that at least maintenance information is used – if no black-box or flight recorder data is available then at least via feedback cases and returned devices.

Examples of listening to users and potential users here are presenting demos for feedback, interviewing test users, and possibly publishing ideas, views, and floating upcoming features via blog posts, news articles, interviews, etc., to get an echo from potential users. In the future, we fully expect AI-based chats to increasingly feature here as a medium for users to get help and manufacturers to gain insight into the paint points users experience. The above list of examples will perhaps all be exercised on the platforms created to support the AI chats.

8.3 Risk management

Risk management is closely linked with product realization, both during initial development and then later as changes to the product need to be made. A particular upcoming topic we see here is the tracking of realized benefits as well as the risks we all hope didn't materialize (see Section 6). Unlike the risks we don't like to see we do want to see benefits materialize, and this may be a small revolution in long-term risk management activities. Artificial intelligence and the slow march forward of interoperable databases, such as those created for adverse events around the globe, may offer us great opportunities for automation of information gathering activities and early-warning mechanisms.

8.4 Usability engineering

Usability engineering, too, is an expected part of the initial development of a product and an implied aspect of managing changes to the product later. No simple rules or guidance exist on when to revisit usability activities (see Section 1.22) beyond the obvious which suggests that when a device or its use changes significantly enough we should perhaps do something. We feel that gradual advent of benefit management, discussed above and in Section 6, is one of the strongest cues for revisiting our understanding of usability around a particular device. The IEC 62366-1 standard does give us the formative and summative usability activities needed to address the use, use environment, and use of a device in standardized way, and these activities should also form the basis of our revisits to usability engineering over the long term. More guidance is perhaps forthcoming in, for example, the form of IEC 62366-3 document on the application of usability engineering to devices based on artificial intelligence.

8.5 Post-market

Post-market surveillance is the ultimate manifestation of listening to the user over the long term. Here knowing all the sales figures, customer feedback, market intelligence, and the advancement of science and the competition around a device, the manufacturer is asked to keep analyzing their understanding of the user and the device. This may involve new usability engineering and clinical evaluation activities that may be tuned to broaden or reaffirm what the manufacturer thinks they know about the user.

All of the above activities are relevant for keeping taps on the use and the users of a device. In many ways there is no need to reinvent the wheels here: the prehistoric or stone-age wheels of the Flintstones will continue to work here just like the neon-striped speeders from Tron or any number of futuristic sci-fi thrillers. Yet, with artificial intelligence we may be imbuing devices with more complex and more powerful features than is immediately apparent to the user.

Chapter 9

Integration with Quality Management

Usability engineering should not just be an isolated gate you pass on the long journey from an idea to a released product, nor should it be an activity you run every once in a while, like a bilge pump keeping your boat afloat. Usability engineering should connect with the other relevant processes of your QMS, as illustrated in Figure 9.1, *and ensure that the correct questions are asked at the right times, and the correct activities are then undertaken.*

One of the cardinal sins of usability engineering in the development of medical devices is that it's an isolated activity perhaps run using a representative sample of the device but otherwise conducted in a vacuum without many inputs or outputs to brag about. The purpose of usability engineering under such circumstances may not be to find flaws or discover better ways for the device to achieve its goals, but to find a plausible rationale for thinking that all is well. The danger of running such a checkbox activity increases if we only consider usability at the point where we have a product and want to get the appropriate stamps on its release papers.

Performing some checkbox-like gate activity on a device is infinitely better than not paying any attention at all to the usability of the device, but it is not what the standard, IEC 62366-1, has in mind. As pointed out earlier in Section 1.10, usability engineering is intended to be a documented process linked to not only the development life cycle of your device, and its risk management, but also your other key quality management activities. The guidance on usability engineering, IEC 62366-2, goes as far as to call a

DOI: 10.4324/9781003645702-9

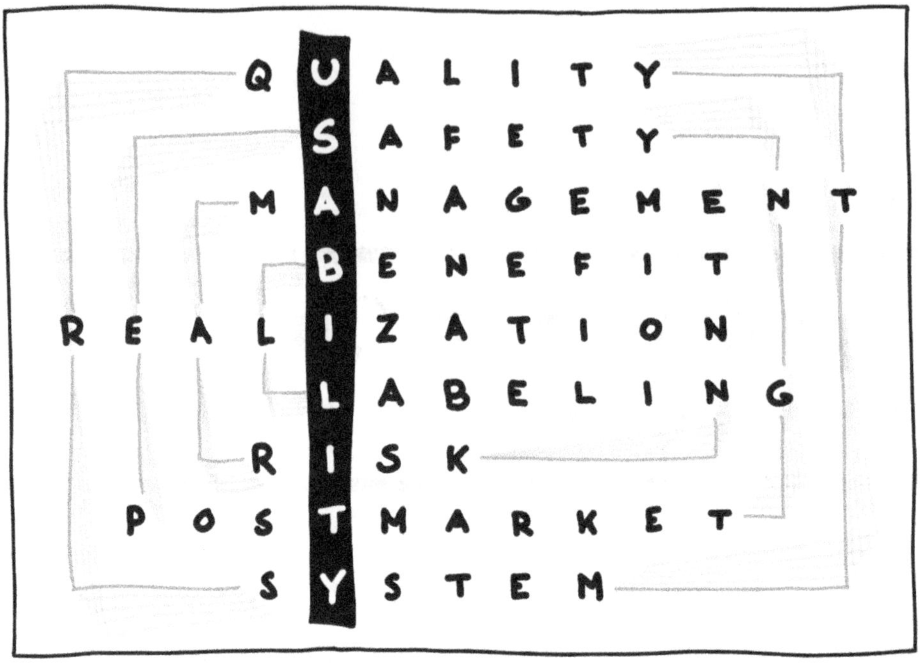

Figure 9.1 Usability may cut across a lot of the activity involving medical-device interfaces.

high level of integration with other engineering, design, and even marketing processes a hallmark of an effective usability engineering project. We wholeheartedly agree and add that also the integration of relevant long-term activities matters here: change control, post-market surveillance, and the perpetual risk management carried out around our released products all benefit from awareness of usability engineering. These links should be planned in a QMS as best suits the practicalities of the organization and the medical device they work with. In the following we'll discuss the linkages from usability engineering to product development, risk management, and some other key quality management activities. On a high level, these linkages are illustrated by Figure 9.2.

In the figure, usability engineering and risk management find common ground where use errors can lead to risks and then perhaps having to be mitigated in the product, its documentation, or the processes used in manufacturing the product. Note also how some of the identified risks will come from outside the realm of usability engineering. The view is necessarily a simplification as risk management cuts across absolutely everything in medical devices, but we feel that the figure provides a useful

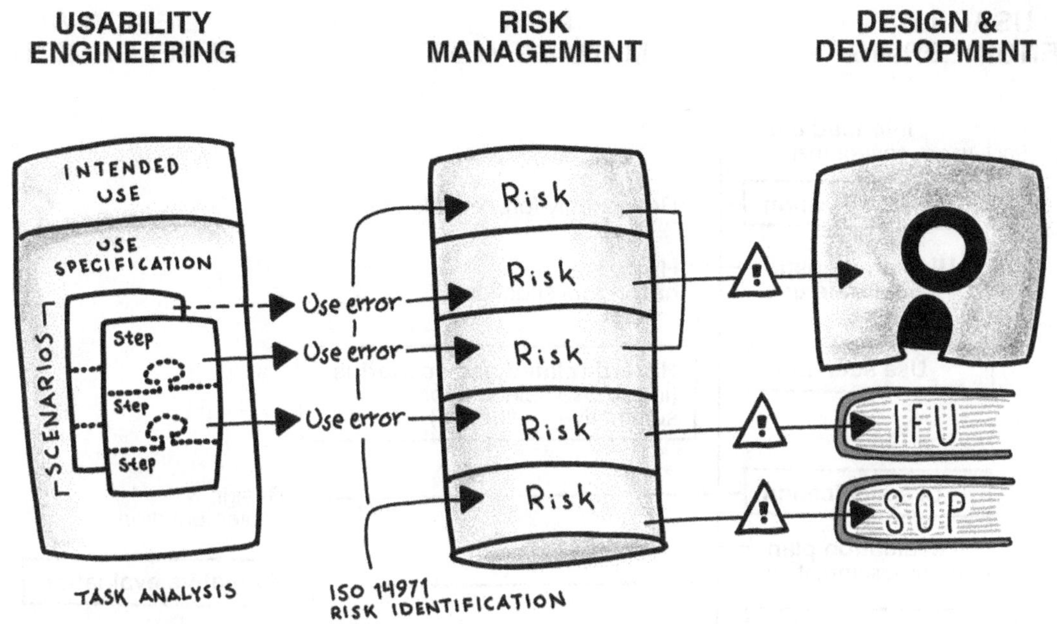

Figure 9.2 Linking scenarios, risks, and results.

overview of the general linkages between the three processes. A more detailed view is given in Figure 9.3 where the key activities of usability engineering are linked with the other two processes.

In the figure, time flows from top to bottom so that we might begin with our understanding of the intended use of the device (including users and use environments) and then proceed through the rest of the activities to finally make an ultimate benefit/risk assessment for the device we developed. The major items enabling key communication across the processes are framed in the figure. In essence, the figure aims to show how each of these three process domains progress on their own, at their own pace, and finally map together via critical phases to develop a single medical device somewhat like what happens in Christopher Nolan's film "Dunkirk" with the three different timelines progressing at different speeds to merge at the climactic end.

Note that the two figures presented here, Figure 9.2 and Figure 9.3 represent two different ways of looking at the links between the processes. The former looks at how the output of one affects the other while the latter looks at how the activities of each could be connected. In both figures we have drawn one-directional arrows to illustrate a point, but in reality all the arrows can occasionally be thought of as two-directional: detected use

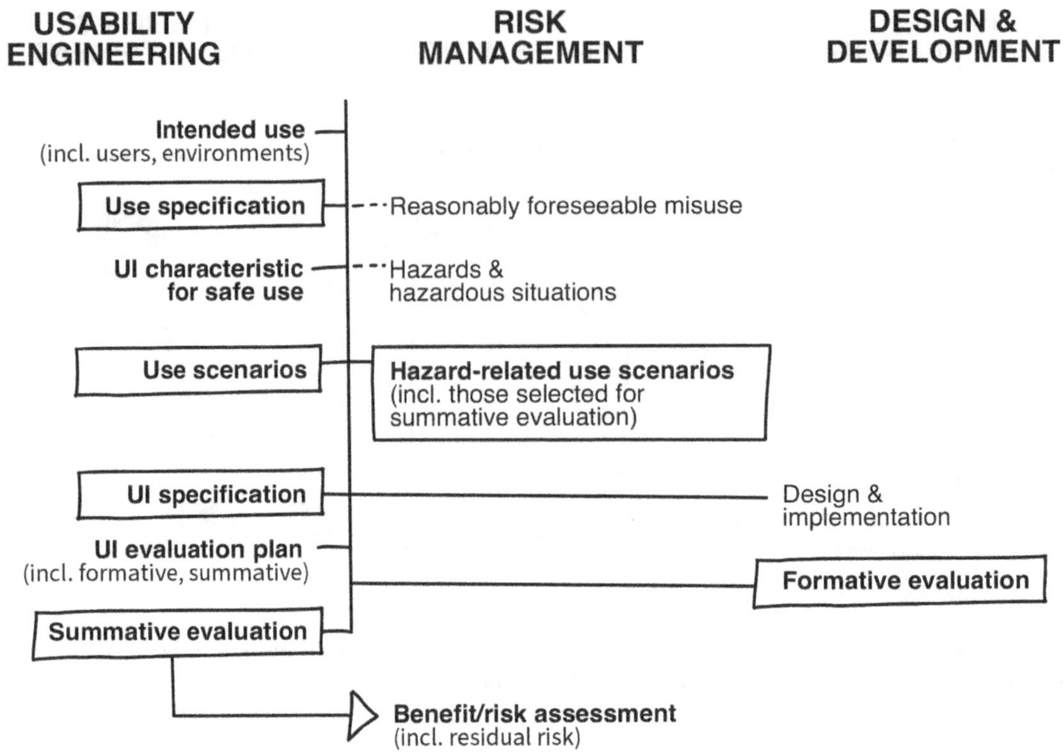

Figure 9.3 Integration between usability, risks, and D&D.

errors affect our understanding of risks, but also the identification of a new risk – or some reported adverse event – may lead to the identification of new use errors and potentially new use scenarios.

In both figures, we gloss over the details of the design and development work to enable a higher-level discussion of the linkages between the domains, but we will address this particular area of overlapping activity in more detail in both Sections 9.1 and 10 (also see Juuso & Pöyhönen 2024). Somewhat controversially, in the figure, we place formative evaluations in the domain of design and development. We do this to highlight the synergetic nature of the almost freestyle formative evaluations and the design and development work that will take place between the UI specification getting inked and the final summative evaluation being performed. We feel that the requirements for formative evaluations from the side of IEC 62366-1 usability engineering are relaxed enough that as much input probably comes to their conduction from the prototyping and staging involved in the design and development work as from any UI specification and UI evaluation plan at the common root of all these activities.

In the following, we'll discuss each domain of activity in turn from design and development to risk management and beyond. What follows is rooted in our own wrestling with the usability engineering standard and how to fit it with the world we happen to live in. You may read this section with a grain of salt, take the elements that hopefully strike a chord with you, and forget about the rest of the musings in this section. After all, each quality management system will be a little different, even if the standards are technically the same for all.

9.1 Link to design and development

As hinted at in the previous section, usability engineering activities and the design and development activities may proceed toe-to-toe with formative evaluations and prototypes guiding development work from both the developer and the user perspectives. The engineers writing the code or bolting some models of physical components together will try to understand the user, but they will nonetheless be interpreting the requirements through their own mindset. The user, on the other hand, rarely has an appreciation of how the product (software or otherwise) needs to be constructed to work with existing materials, components, and even software programming languages. To force the developer and the user to somehow meet each other may thus be beneficial, and in this usability engineering holds a winning hand of cards to play. Figure 9.4 gives one example of how usability evaluations (both formative and summative) might be linked with the progress made on the design and development of the device itself.

In the figure, the frontend represents the interface the user will see and interact with while the backend is the engine room where the developers mostly live. It may be possible to fake functionality and thus commence formative studies with limited prototypes and demos, whereby the user response to developed features can be tested before all the modules are in place to run the feature for real. This way we can do something beyond paper prototyping before the finished device is in place. Remember how in Section 3.7 we discussed different ways of prototyping? This is an example of how that particular tool from our toolbox can be wielded for gains from both usability and implementation points of view. The engine room then incrementally develops to support more formative assessments until finally all code is written and all hardware molded to support the ultimate summative assessment using the final release candidate.

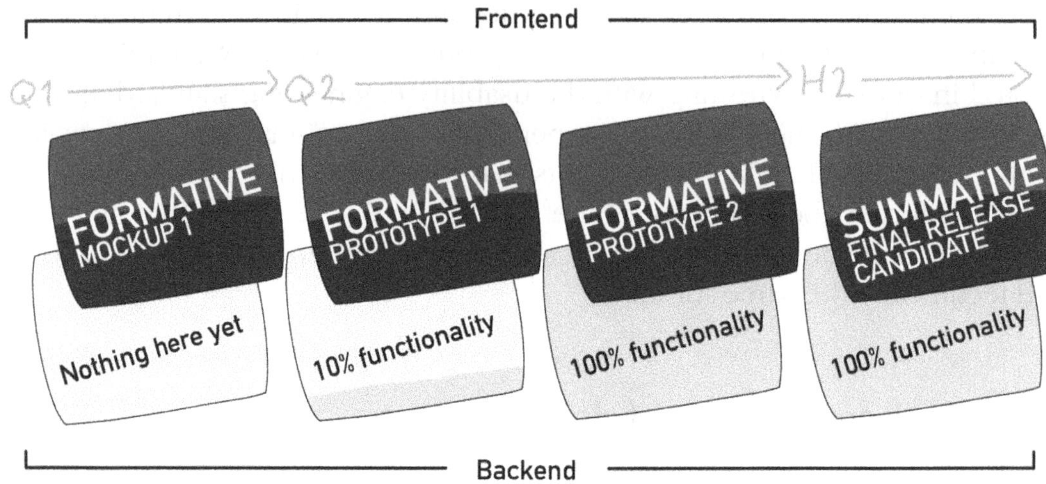

Figure 9.4 Usability evaluations and D&D maturity.

A fairly reoccurring issue with usability engineering activities is that, for one reason or another, the findings of the assessments are not translated into real effects on the product, its documentation, and the relevant product realization processes. The existence of reports is somehow forgotten, their results are ignored, or the suggested corrections waitlisted. Structuring development to coincide and indeed take place in a symbiotic relationship with formative usability activities, will go a long way towards dissuading such disconnects and allowing for proper responses to be understood and enacted. Having the same people work on both sides of the coin, if their competence allows this, may be a good approach, but having the experts from both sides routinely interact with each other may be an even better one. IEC 62366-2, too, recognizes the importance of this crossover point by stating that poorly timed usability activities, while perhaps still providing insight, may not do so at a point in time when this insight can easily be acted on. How an organization wants to structure this dialogue is up to the organization itself, but the spots to hit by the end are set out by the IEC 62366-1 standard (see Section 2).

In the following we'll briefly discuss how usability engineering may be brought to bear during the design and development of a product. Later in this book, in Section 10, we'll go ten steps further and introduce our roadmap model for progressing from research to implementation to release with product concept.

9.1.1 Requirements specification

A famous quote attributed to Henry Ford has the early American automobile innovator quipping that if he had asked his customers what they wanted, they would have said a faster horse. The quote is often showcased as an example of how the users may be incapable of articulating their real needs: instead of asking for a fancy new T-model automobile they might just ask for a faster specimen of the horse they already have. It is no doubt true that users will employ the vocabulary and concepts at their disposal to voice the wishes they may have, but it is also equally probable that the developers will hear the things – vocabulary and concepts – they are used to working with. In the quote, the users asked for a faster horse and that is exactly what Ford delivered to them in the form of a fast automobile.

The purpose of the above story was to point out that usability engineering holds just the right sort of tools to bridge the gap between users and developers. Instead of developers building a finished product and then quickly running it past users, they might use the usability toolbox to talk with users from the very start. At the beginning no dollars have yet been spent and there aren't yet a myriad of design decisions to explain to the user to gauge their interest in the concept. The user may even have useful things to say that may then affect the way the requirements for the product are written. The developer, too, will be able to learn how to speak with the users and what truly matters to the users in the context of the product. If a user then says that the faster horse they want should be able to travel more than 50 miles per day, the developer should use this to design a sufficiently large gas tank into the car – and realize that a car could do many more miles beyond what the user may conceive as physically possible. The concepts used on both sides may not be readily understood by the other side at first, and the mapping between the concepts may take some effort, but over time the differences will grow smaller. It thus makes sense to start working on identifying and bridging over the differences when the costs involved are counted in post-it notes instead of million-dollar prototypes. The right tools to do so are given by usability engineering.

In engineering the pressure will always be on finishing the thing or just getting something done, while usability engineering theory appears to be written from the position that resources, test users, and time will always be abundant. In reality, time is always a constraint no matter what the budget, and a small budget will just make that time seem even shorter. In moments like this it doesn't help to drag our feet, but it may be worth remembering that General Dwight D. Eisenhower considered tasks to require either urgent or thought-out action, never both. Eisenhower thus sorted his to-do list into urgent and important items – it was only later management specialists who turned the two-bin model into a four-bin model and began talking about delegating unimportant urgent items. For us, things are either worth doing or not worth doing, and the latter should just be binned for the landfill.

Over time usability engineering, perhaps boosted by post-market surveillance, may also lead to accumulated knowledge about how users interact with the exact device or the types of devices in general contained in the organization's portfolio. This information may be retained in some form of a practical repository of requirements, questions, and checklists for use by later projects, much like a translation memory might be used by translators as a global or persistent vocabulary. This type of information may even become a useful part of what the standards and regulations ask for as information about previous designs, although it perhaps does not cover everything you may have had to learn from previous designs and projects. The standard does not make any requirements on the use of previous usability information or provide any tips on the topic, so if and how you build such a feedback mechanism is entirely up to you. As a way of recording some of the silent information in the organization this might also be a tool worth having.

9.1.2 Design

If the user is involved during the requirements specification phase (see previous section), it probably makes sense to also keep them somehow in the loop through the design of the product. This can entail the same sort of prototyping and interview activities as was used to arrive at the requirements for the device earlier. We may have more mock-ups to show the users and quiz them on, but we may also want to ask them some well-crafted questions about the design choices we want to make. In most cases, however, the user will not care enough or have the requisite competence to assess some detailed design document so we shouldn't try to run just anything by them. The user is not a rubberstamp.

It should go without saying, but the user involved here during the formative evaluations should not be part of the group of users used later in the summative evaluations. Here during the development phase, you may use employees of your company or other closely affiliated guinea pigs, but if you do, realize that they may not be completely objective and the further they are from your intended user the less likely they will probably be to step into their shoes. Taking a layered approach to usability testing here, perhaps first using your own people in routine exercises and then reaching out to outside subject matter experts once you have something more polished to show them, may make sense. Even then, don't prime your summative user group by involving them in your formative work.

9.1.3 Implementation

During implementation we will have increasingly concrete items to show the user. These will range from simulations to demos and finally prototypes. At each step important feedback could be solicited and increasingly refined insight gained from the user. Here we are not yet talking about summative evaluations so it may not matter that the users, if they remain the same throughout our design and development work, are constantly learning more about us and the device. The partnership that forms here might, thus, be just the right approach to finding the optimal solution to the device we hope to make. However, beware of either side overpowering the other: you may be able to bend one set of paid test users to your will, or they may take you for a long walk in the wrong direction unless you keep your wits about you. Also, the same words of caution on priming your future summative evaluation user group as voiced above apply here.

One special type of testing you may conduct as part of your formative evaluations is A & B testing, where you show two solutions to your users and ask them which they prefer. This type of testing may give you valuable clues to how to meet some design consideration and it will be much easier to conduct here than in a later phase when you can't really do it (i.e. in the summative evaluation when you want to see if a certain version of the medical device floats) or it may be costlier to do as you'd perhaps have to have separate A and B groups of users and see which group liked the one interface which was shown to them better. Perhaps there is no reason why you couldn't do A & B testing until the very end of your formative evaluations, but it is better suited for early tests with prototypes where you can create the variants with negligible cost, but the variants nonetheless have distinctly different characteristics. Running parallel summative evaluations to do A & B not only sounds like an expensive proposition but also more than a little suspicious.

91.4 Testing (incl. verification and validation)

Usability activities such as both formative and summative assessments can certainly be used as part of your overall testing activities. They will serve as a valuable part of the evidence demonstrating that the device you put together works well from the point of view of the users you recruited into the assessments. They will give you confidence in knowing that it all works and perhaps even does something worth doing.

Usability evaluation will not, however, prove that you got what you set out to get by writing your requirements and sketching the design documents. You can't delegate responsibility for ensuring that you have crossed every t and dotted every i to the usability assessments. In other words, verification is up to you. You as the manufacturer will need to verify that your output is prescribed by your input, and that all of your design and development activities are complete.

Similarly, usability evaluations won't prove that your device works as clinically intended. We touched on this lesson already in Section 1.15. For clinical evidence on the efficacy of your device you need the activities instructed in the domains of clinical evaluation and clinical investigations. Your body of users may again be somewhat different than for your summative evaluations and perhaps very different than the friends and family you may have used in your formative evaluations.

Usability activities will have had a great impact on the requirements and designs you drew up. They may provide a part of the story in wrapping up your testing or even tell you what to test with what rigor, but they will not release you from performing unit, integration, and system testing in an adequate manner.

9.1.5 Training

In working with the test users over your design and development phase, you will have learned a lot about your users and how they use the device. Having seen how much (or little) instruction they have needed throughout the process you will now be set up for using that knowledge to shape your training and training materials. Instead of having to guess what the user might need, you already know what one set of test users actually needed to get up and start running while using your device.

The intelligence you have built on your user and the sharpening of the schema for exactly who that user is (e.g., their role, background, skills), will help you now fashion the material needed. This will save you a lot of time developing the material as you can take a flying start to the proceedings. Perhaps you have already started developing the material as the implementation work progressed and perhaps you have even tried out some of that material in organizing your usability evaluations. If the training or the training material act as mitigations in your risk management, you will certainly have done so. In any case, now might be the time to take care of the new translations needed and ensure everything is in place for the mobs to arrive. Usability engineering can help you polish any remaining blemishes in the training package, and also improve on any non-safety related factors of the whole. Beware, however, of cutting away at any of the characteristics you may have relied on as part of the usability engineering activities. Perhaps now is also the time to think about how you plan to address usability as part of making changes to the product or its documentation, if you haven't yet done so.

9.2 Link to risk management

Product development and usability engineering may intuitively feel like they are connected activities, but for anyone coming in from outside of the domain of medical devices, or some other equally regulated field, the connection to risk management may not feel quite as natural. Risk management may seem like a complicated activity best left to a separate group of professionals who know how to color their spreadsheets and run their FMEAs. Identifying, analyzing, and controlling risks certainly call for effort and the right skillset, but it should not be seen as something separate for someone else to do. The information is generated in many different fields from engineering to clinical expertise, and all of it should be used to drive safer devices.

In some ways, product development (including usability engineering) is like a Lego model constructed of block-sized activities, documents, and deliverables while risk management may feel like trying to fit Duplo blocks onto the same model. In calling one set of activities Legos and the other

Duplos, we are not thinking of the intended age groups of these building sets, but rather the idea that both sets of blocks sort of fit together however not completely so. In many ways, risk management cuts through absolutely everything in medical devices from the organization that made the devices, to the process these were made through to the actual finished product. If we look at the quality management system and the product development pipeline constructed within it, then rotate the whole by 90 degrees, what we may see is the risk management system running through it all. Everything is connected, and the right events at any point in the model may lead to effects in risk management, and vice versa. Yet, some parts of the network of processes are more connected than others.

In the above, Figure 9.2 already illustrated how use errors and risks can be thought of as the nucleic acid pairings making this particular strand of DNA possible. Newly discovered use errors should not go into a special usability vault but should instead affect our understanding of the risks associated with the device. How those use errors are then sprinkled across our set of use scenarios should be the next level of the onion inside usability engineering, but this is more of an internal usability engineering matter than a topic of joint interest with risk management. Any new realizations about use scenarios, hazardous situations, hazards, use environments, users, or any other event or stakeholder should feed into risk management as appropriate, but not everything needs to be recorded twice and always analyzed on both sides.

In the following we'll discuss the identification, analysis, and control of usability-related risks, as well as verifying the effectiveness of those controls.

9.2.1 Risk identification

Earlier in this book, Sections 2.3.2 to 2.3.4 discussed the identification of safety-related user interface characteristics, hazards, hazardous situations, and hazard-related use scenarios. All of these are potential topics of joint interest between usability engineering and risk management, but the atomic unit best suited to cross over from one domain to the other is risk itself. Similarly, the identified use errors, use difficulties, and close calls related to the risks are of interest on both sides of the fence. Usability engineering should offer the best tools for identifying what risks may arise out of the use of a device and how to characterize those risks. Risk management will then have the engineering insight, device-specific priorities, and command of the whole of the device to ultimately make the necessary analysis and plans for remedying any issue.

Our suggestion is to let usability engineering activities loose on the device to produce lists of identified and analyzed use errors as well as the right intelligence on the user, use environments, and uses of our device. This output can then be used to better understand the product, and feed back into to the usability activities and design and development activities as relevant.

9.2.2 Risk analysis

Once risks are identified, characterized, and semi-analyzed it should be up to the discipline of ISO 14971 risk management to figure out how to best address those risks – at least until it comes time to verify the usability-slanted actions called for in the mitigation actually do work in practice. We would thus relinquish most of the column space here over to ISO 14971, but we do want to make a few points before we do exactly that.

It is possible that ISO 14971 risk management, as much as it may try, will miss the peculiarities of human users whereas the IEC 62366-1 should, in theory, be finely tuned to analyze this very body of actors. We defused some of this issue by asking usability engineering to not just provide raw information, but to refine this to meaningful knowledge of risks the user may be subjected to. We would expand this from just a description of what could go wrong and how bad the outcome might be, to also suggest mitigations that would reduce the probability and the severity of the harm in each case. It is then up to risk management to assess whether these are appropriate, how the risks relate to the other risks associated with the device, and what should be done to mitigate the risks if anything. This would defuse another major issue with the handover: risk management not being equipped to consider usability appropriately while discussing mitigations of the risks. The last part of the bomb we would defuse lies in the verification of the effectiveness of those mitigations: here too usability engineering must be a part of the consideration.

To illustrate the defusing of the bomb, consider Figure 9.5. The figure illustrates how errors in the product and the use errors in using the product may be intwined. The issue addressed in the figure relates to an automatic emergency defibrillator (AED) failing to administer the electric shock to restart the heart when the victims clothing obstructs proper skin

ID	Failure mode	Hazard	Harm	Initiating cause	P1	P2	Risk control	Verification
1	aborted discharge	delayed therapy / ROSC	brain damage	high trans-thoracic impedance	?	?	"no shock" message displayed	oops.
2	confusing message shown to user	delayed therapy / ROSC	brain damage	abnormal impedance detected	?	?	include all messages in usability evaluations	oops.
3	incorrect use	delayed therapy / ROSC	brain damage	confusing error message	?	?	training? IFU?	oops.

Figure 9.5 Analysis of usability-related risks.

contact. The figure provides three different ways the resulting risk might be recorded, with somewhat different outcomes. As seen in the figure, the issue might be seen as an error in the device causing it not to work (an AED fails to work as intended) or it could be seen as a use error (an AED is not applied appropriately so it can't administer the shock).

In the figure, the same error message ("No shock delivered") is recorded on three separate rows of the risk analysis in three different roles. Row one takes a traditional device-centric view of the issue and considers an error message as sufficient response to the malfunction. Here the view is that the user will then react to fix the issue and somehow just "do better" in the future. On the second row, we consider the situation from the point of view of the user who wanted to save someone's life but is thwarted by an indecipherable error message. Here we suggest that appropriate mitigation might be to subject all such messages to usability evaluations where, we hope, all such kinks would be discovered and dealt with before the device ships. On the third row we take a somewhat heretic view and blame the user for not using the device correctly (which is what actually took place behind row one as well). Even here we note the unfortunate error message to be an initiating cause and arrive at a less-than-perfect remedy of explaining the error message in the instructions for use or as part of some user training.

The point in the figure is to say that risks, and indeed usability, can be approached from several different mindsets, but if we follow the reasoning beyond just blaming the user we will get to a meaningful solution for reducing the risks and improving the safety of our device. In all three cases, the message displayed by the device to the user, "No shock delivered", is undoubtedly correct but it may not be the best information to give the user to overcome the issue they are faced with in the situation. A novice first time user might, for example, benefit more from a message adding a tip to check the skin contact. Regardless of whether the issue here is seen as an error, a use error, or (incorrectly) a user error, the crucial part is analyzing it, recording the risk, and controlling the risk through the appropriate steps in designing, developing, and manufacturing the device as well as instructing its use. With that, we will let you refer to your risk management processes or the ISO 14971 standard for more on risk analysis.

One final note to make here while discussing risk analysis is that both risk management and usability engineering both appreciate the need to perform root cause analysis. This technique is not really instructed by either standard and in fact there are almost as many ways of performing root cause analysis as there are experts using the technique, but the fundamental answer always sought is what is the root cause leading to the issue in the first place? There may be several different root causes, each leading to some risk becoming a reality under their particular circumstances, and if that is the case, you are tasked with finding all the potential causes that matter and that you can somehow mitigate. Out of all the ways to investigate an issue and reduce it to its fundamental cause, perhaps the most powerful method is the "Five whys" -method, so named on account of it often taking you five iterations of answering the question of why something bad happened. Even in the example we gave in Figure 9.5 we can observe how the three rows are actually, in essence, the "three whys" of why the AED use failed – although in this case rows two and three are perhaps alternate answers to why something went wrong on row one.

9.2.3 *Risk controls*

Risk control is again a topic mainly for ISO 14971 risk management, but there is distinct overlap with usability engineering when the mitigations suggested somehow rely on action by the user. IEC 62366-1 Clause 4.1.2, too, echoes ISO 14971 in calling for the manufacturer to perform risk mitigation using one or more of the following three options, in descending priority:

- Inherent safety in design and manufacturing
- Protective measures in the medical device or manufacturing process
- Information for safety and, if appropriate, training to users

Note that inherently safe manufacturing and the training to users were both added along with information for safety in the 2020 amendment to IEC 62366-1. The non-binding annex to the standard, A4.1.2, also calls for a rationale to be documented in selecting a risk control option, but no such request is made by the standard itself. This sounds like a messy request leading to extraneous documentation if carried out for each risk, but no doubt the consideration of applying more than one of the risk control options, or even all of them, should be a part of the thought process during risk management for all risks.

The clause includes a note that conformity with its requirements is checked via the usability engineering file, which is thus to either include or reference the appropriate risk management documentation. Note that the requirement does not mention the product itself or its accompanying documentation as possible references here, although these more than likely are in the loop somehow.

IEC 62366-1 Clause 4.1.3 discusses the use of information for safety in a little more detail. Here it is required that when information for safety is used as a risk control that information is in fact subjected to the usability engineering process – in other words, included in the usability assessment somehow as we suggested with the example in Figure 40. The objective of the assessment is to ensure that such information is:

- Perceivable by the intended users of the medical device
- Understandable by the said users
- Supports the correct use of the medical device

In the above, the users, user profiles, and use environments defined for the medical device are of great concern. The standard then points to IEC 62366-2 as a source of examples on information for safety, and notes that such information may occasionally be required by other documents, including product standards. Finally, note that disregard of information for safety by the user is considered to result in abnormal use (see Section

1.6), and thus effectively outside the core risk control activities of the manufacturer.

9.2.4 Verification of effectiveness

Verification of the effectiveness of risk controls (mitigation) is again a topic best appreciated by ISO 14971 risk management, but to the extent those controls face the user somehow also the usability engineering standard IEC 62366-1 pays keen attention. The standard sees a role for itself throughout device development via helping in the identification of hazards and hazardous situations, developing UI-related risk controls, and, here, evaluating the effectiveness of such controls.

The non-binding annex A5 to the IEC 62366-1 standard rightly laments that there is no reliable way to assign probabilities to use errors. Given this, it may also be problematic to truly assess whether some verification has worked or not. If a risk control failed during the summative evaluations that may be a strike against the chosen control or its implementation – but if the control mostly worked or worked to some extent in decreasing the severity of the risk, which could have been worse without the control in place, did it really fail? The judgment here will depend on the particulars of what was being controlled, how it was being controlled, how else it could have been controlled, and also why there was a risk to be controlled in the first place (see Section 6 for benefit management). A clean bill of health is what we always wish for in assessing verification of risk controls, but that may not be the only acceptable outcome.

Even a complete disaster might be thought of as a relative success, in addition to an abject failure. Imagine, for example, that after hitting an iceberg the Titanic started to take in vast amounts of water but still somehow made it to New York. In this case the compartmentalization and the water-tight bulkheads would have worked adequately enough to save the ship. If everyone had survived this would have been called a success. If, however, even a small portion of the people onboard still drowned in some sections of the ship, the mitigation techniques could not have been thought adequate. In reality 1,500 lives out of 2,200 were lost when the ship sank less than three hours after hitting an iceberg. Even then that was three hours less for people to be subjected to the icy cold waters of the Atlantic, and thus much better odds of survival for those involved. The point here is not to make light of the tragedy of the Titanic or promote a devil-may-care attitude towards judging potential fatalities, only to point out that success and failure are at times judgment calls. For this reason, too, it matters how you have defined the risks, both before and after mitigation, in your risk management, and how you have defined any related acceptance criteria.

9.3 Link to change control

Something carefully put together can be undone by neglect alone. This is also true when that something was instructed by usability engineering. The use of a device may change, the environment around it will almost certainly change over time, and even changes to the device itself may be fast tracked around new usability evaluations. The easy call to make here is that post-market surveillance considers any cues for redoing usability engineering and that any change management processes also consider effects of the change on the usability of the device (see Juuso & Pöyhönen 2024). Not every cue deserves a full salvo of usability engineering activities, but ideally, we would know what events, changes, and signals do.

Even after usability engineering activities are perfectly linked with the rest of our product development and post-market surveillance activities, a tough question remains: How should we actually measure the occurrence, importance, and evolving status of something that didn't show up as an error in our summative evaluations and hasn't been observed in our surveillance data since then either? See no evil, hear no evil? IEC 62366-1 usability engineering does offer tools for assessing the safe use of our device including close calls and standardized summative evaluations over time. How you use these tools, how you assess changes in any of the data, and how you then react is, of course, up to you in the first place.

9.4 Link to post-market surveillance

ISO 13485 quality management expects the manufacturer of a medical device to establish mechanisms for keeping their ear to the ground on how their devices are doing once they leave the factory floor. This includes both being open to receiving and actively seeking customer feedback, and also using the maintenance channels for scouring feedback for possible issues (see Juuso 2022). Similarly, ISO 14971 risk management is expected to have a follow-up and monitoring process in place to ensure that changes in risks are observed and addressed as is appropriate.

The standard, IEC 62366-1, has similar expectations on observing changes in the usability of the medical-device user interface, but it does not call for a separate usability engineering thread to be constantly maintained in active use. Instead, it looks towards risk management and points out that usability may continue to be a valid consideration in monitoring risks and reacting to them appropriately. In other words, don't spawn yet another process to keep constantly running, but link your risk management and product maintenance operations to usability as is appropriate. Invoke the usability engineering process from these other processes when and where it is beneficial, for example, when investigating use errors and deciding how to address user interface changes in the product.

Unlike the standard itself, the separate guidance on the standard, IEC 62366-2, ventures into post-market surveillance itself in its Clause 19. The guidance here speaks of reacting to problem reports from the users, soliciting more information from users to investigate issues via specific sample questions, performing root-cause analysis, and addressing the issues appropriately. In such discussion the guidance lands firmly in the domain of ISO 13485 customer feedback handling (see Juuso 2022), but it somewhat confusingly calls this 'post-production review and analysis' (IEC 62366-2 Clause 19). Today, we might more accurately discuss post-market surveillance, instead of just the narrow domain of complaint handling.

9.5 Other quality management processes

In quality management just about everything is connected with information propagating and actions escalating through the network of processes as is needed in each case. Above we already addressed the primary linkages of usability engineering in ISO 13485 quality management, but other linkages may be appropriate in your case if, for example, you subcontract usability activities, purchase product directly affecting the usability of your own product, or see a heightened need to address usability in your feedback handling activities or CAPA. In the next section we will look at how usability engineering can in fact inform the entire path you take from an initial idea or a concept to a finished medical device.

Chapter 10

The Metro Map

Ticket to Ride is a popular board game where players get to build railway empires spanning continents by laying track from one city to the next ahead of their competition. Just like in the game, in medical devices it may be bewildering to figure out where to start from and what work to tackle in which order. This section hopes to provide a roadmap for such work, as depicted in Figure 10.1. *In the following discussion we will go through each zone of the map from an initial idea to a finished product and beyond.*

It turns out that the railroad analogy is appropriate in more ways than one to our discussion here in this book. As a metaphor for navigating from place A to place B it does a great job of driving home the fact that there are more waypoints on that trek than just a home and a destination. Instead, we must plot our course through a potential maze of activities to get to where we want to go. The map gives structure and a sense of linearity to our path, but it also makes it possible to backtrack and revisit activities represented by stations as forgotten aspects or latent issues are detected in any activity previously performed. The real genius of the map is that it allows us to focus on the job at hand at each leg of the journey and, for the most part, leave worrying about future stops for later as we come to them. Following the laid tracks maintains the relationships between activities and helps us remember to consider revising activities at each station we then re-encounter as we resume our drive towards the destination. Occasionally gazing left and right while whizzing along a particular track will also let us consider the other activities that may be taking place simultaneously and that may have some effect on our train, or vice versa.

DOI: 10.4324/9781003645702-10

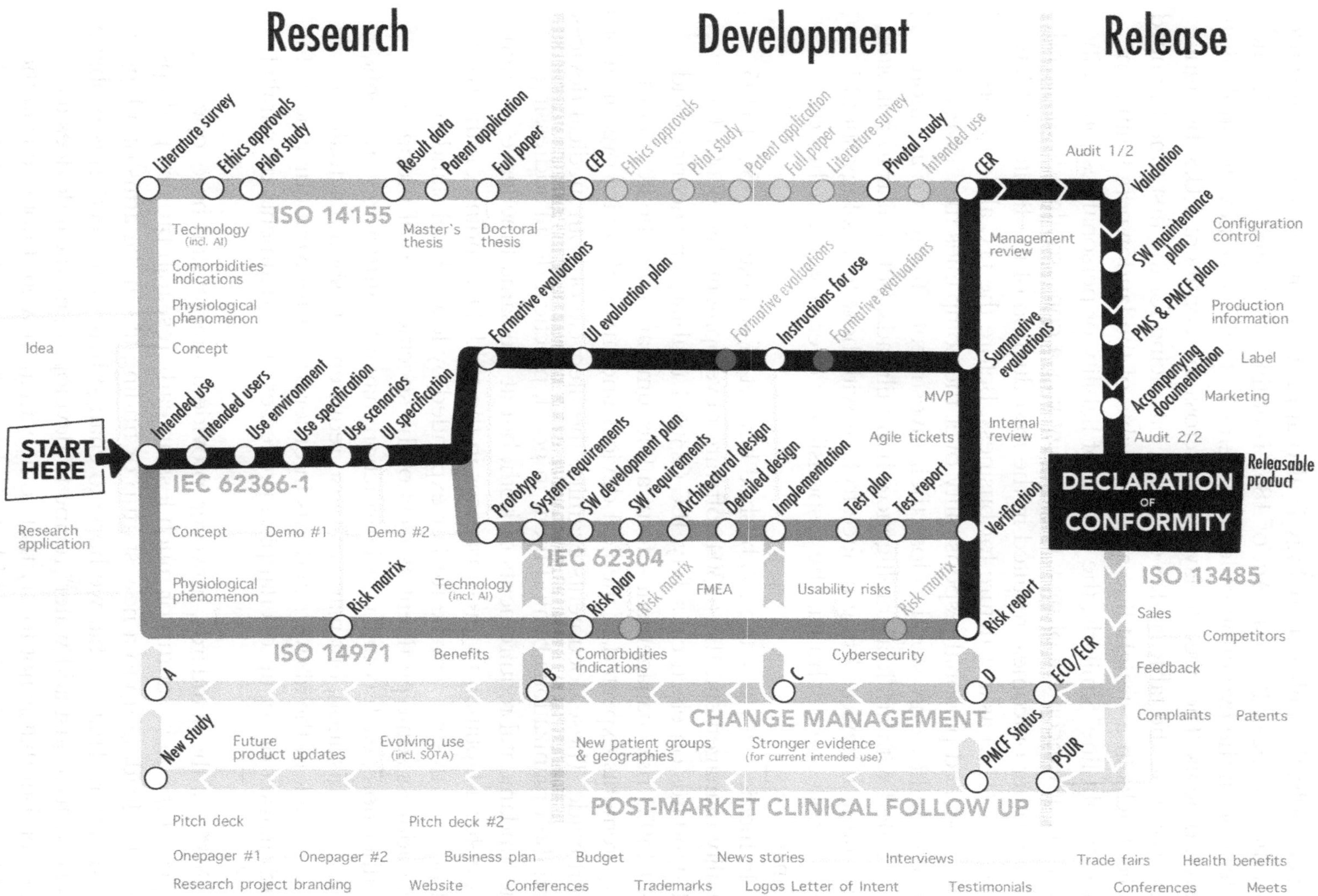

Figure 10.1 A train map from concept to product and beyond.

Secondly, the railroad analogy is also appropriate in the surprising sense that the American Railroad Act of 1862 is at times reported as a significant milestone in the evolution of legal systems in the face of new technological developments. The need to lay tracks across the continental U.S. brought new issues to the table in, for example, balancing for the first time the need to lay tracks in a straight line and the rights of the landowners falling on that straight line. Similarly, the notion of liability or personal responsibility over the advancement of new technology was a concern already then and, among other outcomes, resulted in the introduction of the limited liability company (LLC) as a structure for business. Before the LLC all investors had unlimited personal responsibility over the companies they invested in, after it they only stood to lose their investments. Today, we are more concerned with artificial intelligence than railroads, but similar to railroads, also medical devices bring us face to face with how technology should be controlled and who has the responsibility over how it touches on human lives.

In medical devices, the established key set of standards – ISO 13485 quality management, IEC 62304 software development, IEC 60601 family of standards on electrical safety, IEC 62366-1 usability engineering, and ISO 14971 risk management – form a fair summary over the key set of requirements expected of medical devices and medical-device organizations worldwide. These documents examine the lifecycle of both medical devices and the organizations behind them in good, practical detail addressing the fundamental requirements of both, but at the same time leave enough room for adaptation to circumstances in the wild. Depending on the type of product this set may be further expanded by lower-level standards that address some pertinent aspects of the specific types of products. Furthermore, each country or jurisdiction may add further requirements through legislation, but many of these requirements should actually be traced back to the above standards in order to not reinvent our wheels for each market or set up overlapping processes and documents on parallel tracks.

None of the standards, however, provide a simple 'start here' opening square for a new startup or much less an early research project, perhaps one day leading up to a new startup or finding life in the arms of a bigger corporation. That is what we hope to change here. This section is probably as close to a "just add water" recipe to developing medical devices as we can get, but that is not to say that some variation would not occur in the wild. The roadmap presented here is, however, the only one of its kind we have seen presented for medical devices. Except for a few stations (see discussion below), all the stations are firmly rooted in standards. The arrangement of the stations, however, is something we haven't seen anyone else attempt. We hope this map will be useful to you as you attempt to fit all the required activities together for the first time or discover new optimizations to make in the puzzle you have inherited.

Throughout the writing of this trilogy of books we have taken ISO 13485 quality management as the backbone to all medical-device business. The standard, its underlying principles, and the principles required by the many regulatory frameworks around the world making use of the standard have all instructed our discussions in the books. We have made no secret of our hope to make it all fit, and look for a practical model combining all these elements into a single framework of thought. All of the previous discussions have taken place from some specific point of view: from the overall quality management system of an organization in our first book (Routledge 2022), from the software development lifecycle as part of the QMS in our second book (Routledge 2024), and up to now in this book from the point of view of usability engineering. This is where we connect these individual tracks into a practical model.

A new academic research project investigating a loosely understood physical phenomenon that might lead to a medical device of some sort, a startup looking to commercialize a single device idea, and an established medical-device company looking into evolving a device family all have different starting points. They may have vastly different resources available to them and the approach they take to processes may seem drastically different. They will, however, all want to navigate the requirements as predictably as possible and are all allergic to having to redo work should their path progress to later phases and arrive at a flawed medical device. The question on everyone's lips thus is what should we be doing here and now so that we get to progress to the next step? Or, as the question may occasionally be, should we in fact proceed to the next step?

We have shaped our answer to this need as something we rather pompously call 'The Metro Map'. The reason we decided to discuss this model here in our third book is not simply because we have by now discussed all the parts going into that model, but because usability engineering actually instructs the very first steps along that path. Usability engineering is occasionally seen as the checkered line to cross just before the finish line. Here we argue that it can be used to bring clarity and a sense of certainty already to your very first steps. When in Section 1.17 we said that usability engineering is like the pace car in a NASCAR or Formula 1 race we meant it. The usability engineering standard actually gives us tools and structure to investigate all the key topics we are interested in at the beginning of our product development project whether that starts from a research idea or a device concept. The standard doesn't only give us the moving pieces we should be interested in (e.g., users, use environments, and uses) but also the documents we should record our investigation into (e.g., user profiles, use environment profiles, and task analysis). All of the key requirements are there in the usability engineering standard and the three other key standards we have discussed, but no one has put the requirements into an executable roadmap of activities. That is until now.

In this section we introduce you to our grand unified theory, the one roadmap to give structure to all our medical-device development activities so that at each stage we focus on the right activities and also record those in a way that is readily usable at the next stage. The objective is to, at each phase, dive deep into the topics we should be interested in at that stage – both because we need that information to know if we want to proceed to the next phase, but also because we are actually at a great advantage to look into those very topics at that particular stage without the added burden of all the activities of the next steps clouding our thoughts.

In medical devices it's not rare to see that the plan for doing something has been approved 30 minutes before the report of that activity has been approved. This should not be the case, yet it too often is. Some might feel that since hindsight is 20:20 we should only wrap things up at the very end when we know everything. The standards too expect us to keep many things up to date, but this shouldn't equate to everything always being in a constant state of change. We would argue that the very beginning of the process to developing a new device is actually when you have the greatest potential of focusing on the essence of what you want to develop. This is when you are perhaps lacking in understanding of your user and the medical phenomenon you hope to tap into, but you also have – not just the greatest need – but also the greatest freedom to learn and study these aspects. You don't yet have to distract your work with packing, labeling, and the post-market for example, but can instead focus entirely on what you know and what you want to find out next. Similarly, you won't want to bet your house at the very beginning if you can avoid it, and you won't want to blow all your resources on a project early only to discover that you want to abort the activity before the next phase. These realizations are a big motivation behind the model we present here: to provide focus on the activities at hand and certainty in knowing what those activities lead to next.

We divide the model into three phases (shown as zones on the map): research, development, and release. We then further divide these phases into tracks mostly running left to right on our map based on the aspects they address: usability, clinical evidence, risks, software development, and final release. We also add a few feedback loops – change management and post-market clinical follow up – that will get discussed a little later during the discussion of the ever after (see Section 10.5). The discussion in this section is split into subsections accordingly. So, let's get going.

10.1 Start here

Every story has to have a beginning, and this is that for our roadmap. Technically 'Start here' is not one of the main phases of the model, but it probably is where you are right now as you consider taking a plunge in the pool we have constructed here. So please excuse us for adding an unbilled, cameo appearance for the starting square here.

At the beginning of our R&D project, we probably have more questions than solutions rattling around in our heads. The empty page in front of us, or the weight of the great unknown and the mountain or work ahead may seem daunting at square one. This is like standing in the Go square of Monopoly, as a miniature-sized figurine, and seeing the outlines of the whole future city before us and feeling perhaps a little bewildered by the choice of strategies for the game ahead. Nothing is yet done, but everything is possible. To add to the uncertainty, at the beginning we may know some of the things we will need to solve, but probably not all of them. The good news is that even if our knowledge is still limited here, we also aren't yet preoccupied by the noise of everything else we will need to do or the many design decisions we will eventually have to make and that will start to increasingly affect our later choices. The freedom of square one should thus enable us to focus most on what we most want to achieve. This is the time when we get to exercise a clear focus on the intended use, users, and environments of our planned device. Cutting away all the extra noise and forgetting for a moment all the interconnected moving parts that will come in later as we formalize requirements and risk management, for example, we now get to sketch the broad strokes for our whole project.

In a regulated domain like medical devices, we may also know that we will ultimately need to generate a mountain of paperwork in order for a finished product to emerge, but we may feel like that is still someone else's problem. This is fair in the sense that we can't ever focus on everything all at the same time. Similarly, focusing too much on the form will just formally get us the thing we wanted to get, but it might not lead to the thing we should have gotten. By this we mean that research will always include fact finding, a survey of what is already known, but it should also include an element of surprise and freedom to innovate if we want to do more than just incrementally move the needle. Our message here is that the form and the content don't have to be at odds with each other if we take a measured, piecewise approach to it all.

You may have landed in 'square one' with an approved funding plan for a research project or you may just have a bit of stolen time to take a stab at a new idea, but in all cases if your end goal is a medical device, you had better start considering the expected requirements as soon as possible. If you are rooted in academia, you may feel that medical-device standards don't apply to you as they are something for the businesspeople to observe when punching out the final medical devices from their factory. You instead want to concentrate on the science and not be sidetracked with business considerations. Standards, however, say as much about the world around your idea as understanding the doctors, nurses, and patients that may be using your device. Standards, regulations, and the laws of nature all set up the system your device will need to fit into at the end as much as human interactions will. To close our eyes to one of these areas is to sail blindfolded. Instead, we would softly argue that you are not Christopher Columbus and this is not the sea at the edge of the known world. The ports on the way to a finished device are shown by the map we have drawn out before you.

So, with that let's get cracking. For real this time.

10.2 Research

Research is typically, but not exclusively, the domain of universities and academic institutions where researchers become tickled – usually the more so the more disheveled they appear – by an observation or a theory they want to investigate further. After hitting the academic lottery by having written a successful grant application they throw themselves into understanding every possible facet and minutely specific detail that might be involved in scratching that itch. The objective is to challenge the idea in every conceivable way, including by layers of peer review, and if the idea stands all this testing conclude that we may have something. The outcome of this research may be some revolutionary new idea such as the invention of the atomic bomb, the sequencing of the human genome, or the idea of encoding all information as digital bits. Depending on the research topic the outcome may also be somewhat more down to earth, and occasionally the result may also be negative in the sense that we want to bury the whole idea 6 miles underneath the Earth's crust. In research, however, also negative results may matter a great deal.

The epitome of a researcher may be Albert Einstein who not only had a hand to play in the harnessing of nuclear energy but is also often quoted as saying that if he had an hour to save the world, he would spend 55 minutes of that defining the problem and then the remaining 5 minutes solving the problem. This is all well and good, but the quote says nothing about actually implementing the solution before the world ends. This is not just a funny anecdote to point out, it is also somewhat indicative of the strengths and weaknesses of academic research. Poking holes in a theory to see if it has merit is definitely a valuable endeavor, but unless we are trying to make Swiss cheese the holes by themselves are not enough. What we really want to know is what all the holes mean, and whether there is sufficient merit left in the block of cheese after all that dancing around. If the puddled, bloody mess left at the end is the new heavyweight boxing champion great, but if not, it's time to let the contender go back home with Adrian and perhaps return another day.

We love academia, and we speak of extensive personal experience spanning over a quarter of a century spent in that environment, but the anecdote is indicative of the stereotypical academic mindset: the definition of done in academia looks at a solution being formulated, at most a scientific paper being published on it, but it does not often carry over to any real-world product being devised to actually administer the solution. This is as well and should be, as everyone should focus on their strong suits, and academia is great at dreaming up new solutions, challenging them, and taking development through its early days. The problem, however, comes when the second wave team, the developers and business folks, come in and deem the results from the previous wave as something they don't understand, something that is not mature enough, or as something they would have to completely redo to stand a chance of meeting any outside requirements such as those set by medical-device standards. This may be a death blow to many a good project that then never progresses from research to development, and thus fall short of anything released into the wild. Unless the results are truly mind blowing, any shortcomings in the provided result set may kill off any future development. This is what we hope to change with our model: to ensure the right things are investigated and recorded in the right way. We won't insist on doing everything upfront from concept to regulatory approvals and a finished product, but we will help you know enough of the needs in the next phase to comfortably reach that watershed moment between research and development. In other words, the ampersand in R&D. Developing a viable product is teamwork and it is a relay race, so it makes all the sense in the world to ensure that the current runner knows what is needed to progress to the next leg and the next runner knows that they will be handed a relay baton instead of a dirty jockstrap for example.

In the following we'll discuss the key steps involved in the research phase of our metro map. This is where we are getting to know the pieces of the puzzle we want to tackle, and do our best to mold them into pieces that might fit together and produce something great. In the discussion we look at three main tracks: usability, clinical evidence, and risks.

10.2.1 Usability

Our roadmap lists several activities that are topical during the research phase, even if the terminology itself is new to most research projects not well-versed in usability. The activities can be roughly divided into four groups:

- Intended use, users, and environments
- Use specification and scenarios
- UI specification, demos, and prototype
- Formative evaluations

In the following we'll discuss each activity in turn.

Intended use, users, and environments

Understanding the user is self-evidently near the core of all development activities. Even if the very core itself lies in some new technological advancement that, for example, allows us to turn scrap metal into 24-karat gold or predict a patient's entire medical future based on a single EEG reading the technology only has value as that value is measured through the user and the patient. The standards and the regulations too naturally put the patient on a pedestal and act a bit like that grandmother you may have who keeps going on about how her grandson will do great things in life. At least to her bridge group. The point here is that the user/patient is our sun and our moon, and our project will either live by or die by their whimsy – and the benefits reaped by them. It thus makes the greatest sense to get serious about understanding these stakeholders as fully and correctly as possible, and as soon as possible. Getting to grips with these topics is the bread and butter of usability engineering.

In Section 3 we discussed the various tools the usability engineering standard and its associated guidance, IEC 62366-1 and IEC 62366-2, see as helpful in methodically understanding the user. These include background user research (Section 3.1) and, perhaps as an extension of that, user interviews and observation (Sections 3.2 and 3.3). These tools provide us with a loose framework for getting a handle on who our user is, how they are expected to interact with our device and where.

Section 4 went into more detail and provided schemas to use as templates when recording our understanding of the above. Intended use (Section 4.1.1), user profiles (Section 4.1.2), and use environments (Section 4.1.3) all deserve our attention at this early stage and all deserve to be documented in way readily understood by the next phases of usability engineering. Intended use is something that can perhaps be appropriately documented, at least for now, based on the physical phenomenon or technology our project works with. For the rest, the tools discussed should at the very least provide a welcome prosthetic to our intuition. Following the examples provided by those sections not only means that we use the correct concepts to structure our records, but also that we pay attention to all the aspects expected by usability engineering and our subsequent development steps. It also means that we don't have to wonder what might be of interest to us in analyzing users, environments, and uses as the boxes to fill are all laid out to us for the first time. You may, of course, discover other boxes to fill, but this is the standard set expected in general.

Speaking from personal experience, writing an intended use (or intended purpose as it is known in Europe) may seem like a difficult task at the beginning of a research project. The intended use may seem like it should be the last thing you write after ensuring that all the things you wanted to build work, are in the box, and can be shipped to real users. It may even seem like a superfluous piece of literature the real users will never actually read. Now, it is true that the intended use must match with your finished product at the end, but it is also the anchor point through which regulations are applied to your product and through which you yourself know how to address the requirements stated in standards for various types of devices throughout development. You should therefore have a good grasp of at least what a draft of the intended use could look like as soon as possible even if you then revise it throughout your research phase. In terms of what goes into an intended use you should be able to lean on your understanding of the phenomenon, the specific technology you use, your intended users, intended use environment, and the intended benefits of the device you are developing. In other words, these are all properties your usability engineering efforts will inform.

At this stage the activity in mapping out the above information may be a quite mechanical operation. In forcing yourself to dissect the users and use environments using the forms presented earlier in this book you will begin to understand each of these more closely than you perhaps expected. You will start to notice similarities and dissimilarities between the groups of users and potential use environments, and you will have begun to choose how you want to limit your target user group and the habitat for your innovation be that a device or a technology somehow to be used by future devices. Never ever will it be possible to do everything and for everyone. Saying you can do everything will be heard as you probably won't do anything by those holding your purse strings. So, you won't fool them, and you shouldn't fool yourself here either. Use the structure given by usability engineering to better understand what you are doing, and document that knowledge in a way facilitating the next steps.

Use specification and scenarios

Speaking of next steps, having a good idea of who your users are, where they will use your device, and to what purpose is a great start. Then knowing exactly how they interact with your device is the next, deeper end of the same pool. That "device" may be a whole, self-contained object or it may just be a particularly spiffy algorithm or other component that allows another device to reach a higher level of intelligence somehow. So, don't succumb to the easy excuse here that you can't yet concern yourself with such questions about the end use as you don't yet know the plastic shell around your device or the cardboard box it may ship in. Your device is, quite simply, the entirety of the thing you want to develop for medical devices. If that entirety is well-defined and closely delimited then congratulations, you have just made your work here that much easier. The devil is in the details of that device, of course, and whether you see that devil as Al Pacino in "The Devil's Advocate" or Jack Nicholson in "Batman" this is where that moonlight tango really begins.

The usability engineering standard positions task analysis (Section 3.4) as the one designated solution to sketching out how a user performs tasks with your device. The outcome of this is a list of scenarios, which as Section 1.11 explained consist of tasks and the various properties and goals assigned with the scenario. It also points to PCA analysis (Section 3.5) as a way of addressing issues with the user's perception, cognition, and ability to take action within these tasks. We provide a simple schema outlining the related expectations in Section 4.1.4. The standard then sees the ultimate format of all this accumulated information take the shape of a use specification (Section 4.1.5). This specification is essentially the coversheet to all the information we have accumulated up to now, or the pretty ribbon around the intelligence package we have put together for the development of our device.

Usability engineering as a mode of study is mostly interested in the user interface, but in reality, all the information we have collected here is relevant for all other aspects of the device as well. In developing a good device, we don't just want to understand use for the sake of creating a user interface – we also want to understand the use itself to facilitate it. Usability engineering, thus, is a tool to understanding much more than just the interface it ostensibly culminates in. Once we know, we know, and we don't just want to use that knowledge to engineer the interface without also engineering the right bits to back it up. That would be like John McClane working hard to wrap his head around the terrorists' plot in Die Hard, and once he's figured it all out just going home without lifting a finger to stop them. Or Columbo, Poirot, or any number of private investigators gathering up all the murder suspects at the end of their investigation only to shake their hands and send them on their way without that 'one more thing'. Such plots frequently get explained in synopses as heroes having to solve puzzles, but no one – except perhaps Albert Einstein – would be happy with just the solution and no effect.

UI specification, demos, and prototype

Above we already lifted the curtains on the engine room behind the user interface, and how the same usability engineering documentation created to instruct the development of the interface also helps us organize that engine room. Here the work could thus be split between say frontend and backend development, with both going off to build the part expected of them. This would be just like a railway tunnel through the mountains could be dug from both ends with an aim to meet in the middle.

In reality, many times you will need to develop incremental demos and an ultimate prototype (see Section 3.7) for your interface work to be understood and assessed while work is progressing. You will have various stakeholders for the project, and they will all want to see progress being made in a direction they can understand and get behind. The need to 'show and tell' will exist in many research projects whether or not they have ever heard of usability engineering. These demos can take the form of post-it notes, powerpoint mockups or whatever else happens to make sense in the context of the project, but the crucial thing here is that they all get documented along the correct lines and feed into the usability engineering package we put together. This doesn't mean that you must document everything in triplicate form and have audio recorders running throughout your meetings, but instead that should get you off on the right foot documenting the demos and always consider any changes between the demos against the same framework of boxes you filled in earlier. This way you will not just generate more noise but instead enrich what you know about the use, the user, and the use environment.

Ultimately, at the end of the research zone tracks, before entering the next phase of implementation, you may have a prototype: something which after all the demos does a good job of simulating the whole you wanted to create in the project. No special meaning is assigned to a prototype versus a demo by the standard, in fact these are not mentioned by it, but both are a part of research and development. Intuitively a prototype is a fairly comprehensive demo that you are happy showing off as the end result of your research. It will not be the final product you will sell as a medical device, but it may be on its way to achieving that status.

Formative evaluations

Usability engineering basically consists of two big-ticket evaluation exercises: formative evaluations to run all the way through development (Section 2.4.1) and summative evaluations to assess the final outcome (Section 2.4.2). A research project could document all experiments with demos as formative evaluations, and it could even perform a summative assessment with the final prototype. As long as the formative evaluations are designed to look at a worthwhile aspect of the device, they will all provide valuable information for future steps. Similarly, the summative assessment could provide a good comprehensive overview of what had been done. However, the summative evaluation most likely could not be done at this stage with a final representative device being used by a representative group of users, and thus any summative evaluation now done would probably be – by instruction of the IEC 62366-1 standard – downgraded to a formative evaluation when the device evolved in the implementation phase. This is not to say that the effort would be lost or would have been in vain, only that we still have layers of the onion to peel away before we can be confident that a summative evaluation would stick.

In any case, formative evaluations should be a part of capturing what takes place during the research phase of our work and ensuring that work leads to enriching the set of information we have. This information is, in all likelihood, already generated during the research phase, but it may not be as balanced, controlled, or structured as it should for our use both here in the research phase and later as we move ahead with development.

10.2.2 Clinical evidence

Usability engineering may seem foreign to the average academic research project, but clinical evidence probably will not. The notion of what is clinical evidence, however, may elude some projects. All evidence generated during the research project will be interesting going forward, but not all of it may be useful in making a clinical case for the device we are developing. Here are a few pitfalls that erode the value of the data we might have collected during early research:

- The data is from a different demographic than what is ultimately seen as the intended user/patient population

- The data doesn't adequately cover the full breadth and depth of the intended user/patient population
- The data was collected using a different device or a prototype device that is markedly different than the final device
- The evaluations were performed using a protype device that no longer matches with the device being developed
- The data is from a setting quite different to the intended use context, clinical or otherwise
- The clinical evaluations were not instructed by ISO 14155

In other words, data gathered from the indigenous peoples of Java may be tremendously exciting in the academic sense, but it will probably not convince the FDA that the data matches well with the demographic makeup of the U.S. Similarly proving something with healthy college-age volunteers from Ivy League schools in the U.S. probably doesn't translate directly to the demographics of Inari in Northern Lapland. In research any of these groups may be a valid subject group, but for a medical device they may be very restrictive in its area of application.

If the above stung a little, the good news here is that all the scientific tools of analysis will probably already be familiar to the researchers working on the project. A medical device still fits into the domain of medicine, and the evidence we seek is, in part, to be derived from our understanding of the science and the state of the art in medicine. We don't say this to make light of clinical evaluations, only to encourage everyone to trust the scientific process they know, if they know it, and otherwise learn it. Perhaps the only lesson for researchers with a scientific background to learn here is that in addition to doing science also the business side of the equation should be woven in, at the very least in terms of considering demographics of the data – and in tending to patenting before it is too late. That is a low-hanging fruit, if there ever was one.

The activities here can be roughly divided into four groups:

- Literature survey
- Ethics approvals and pilot studies
- Result data
- Patent applications
- Scientific publishing

In the following we'll discuss each activity in turn.

Literature survey

An academic research project will have a natural drive to understand a phenomenon, measure it, and act as seen appropriate. Performing a literature survey in this context, and against a body of scientific publications perhaps reporting on earlier similar endeavors, is a great starting point. The amount of information we can learn through such an exercise, and the tools to use in assessing such prior information, will provide a genuine boost to our device development activities – and as we learned in Section 3.1, the tools are not markedly different for medical devices than in any area of scientific study.

A research project may, of course, also stem from other backgrounds besides an academic one from the intersection of medicine and engineering. There is no right way or wrong way to start investigating a concept, but sooner or later the model used to assess that research will be the one from academia so it's worth getting at least familiar with those expectations.

The topics we raised on the usability track above will, in part, inform how we point our floodlights here during the literature review. The same phenomenon, technology, users, environments, and uses act as sighting aims for us here. Looking at similar devices, previous generations of any technology we are developing, and any real-world use as well as theoretical study is of interest to us. If we are dealing with a revision of an already existing product, we will want to look at the track record of that product. All this information can be recorded in whatever format makes sense for our work but knowing that the work may need to be cross-referenced later by work on the other tracks or by future iterations of literature studies will help us fashion a meaningful structure to both the documents created and our overall activity of creating them. You will already know many of the key information sources to query based on your field and the novelty of your device, but if the domain of medical devices is new to you, you may consider some of the following:

- Manufacturer and User Facility Device Experience (MAUDE) by the US FDA
- Database of Adverse Event Notifications (DAEN) by Australia's TGA
- EUDAMED by the European Union
- Database of 510(k) Premarket Notifications by the US FDA

You can find links for all of the above, and for further national databases kept of medical devices and adverse events, by searching online. Also consider signing up for newsletters from these organizations to hear of potentially relevant recalls and other events of interest. Finally, if you are not an expert in conducting literature searches, and you would like to avoid potential pitfalls in the domain of medical devices, you may wish to consult a medical-device consultancy that has compiled such surveys in the past.

The perhaps surprising realization here is that a literature study is not a "once off" thing but will instead end up being a constant part of your product lifecycle activities from here on out. Once the concept becomes a finished product and is placed on the market, the same sort of follow-up will take place as part of the periodic post-market surveillance expected of us for all our products. The literature survey may become a part of a periodic safety update report, or its status may just be referenced by such a review, but nonetheless all of the same types of information sources will need to be polled from here to eternity – if not by you, then the organization ultimately taking custody of your device.

Ethics approvals and pilot studies

In some cases, scouring the realms of written information may be enough to build your clinical evidence, but in others you will also need to conduct clinical studies yourself. Performing an interventional study somehow affecting the study subjects is naturally a much bigger mountain to climb than a register study based on already captured data of real patients. Data privacy will be a factor in just about every case, but in the latter, it may form the majority of the hurdles you need to cross. Ivan Reitman may have gotten the permission to separate Arnold Schwarzenegger and Danny DeVito at birth, but you are practically guaranteed to not get the same treatment – something which is even reflected in the sequel to these adventures where Schwarzenegger resorts to impregnating himself with a human embryo as he is frustrated by the proper process.

Whimsical film references aside, the ease with which you obtain ethical approvals to conduct the studies you want will necessarily depend on the type of study you want to conduct. Much also may depend on your association with a suitable research site. If you already have project members working in suitable clinical settings, a clinician-led study may be simple to set up. If you reside on the next layer out as a member of a research team at a university or a university hospital, you may find that your organization already has a fairly straightforward procedure in place for applying and performing clinical studies. If you find yourself further out but still looking into the hospital setting you may find that biobanks and other outward-facing hospital entities may provide the solutions you are looking for.

In all the above cases obtaining the necessary approvals will likely require you to submit information on what it is you wish to study, why, how, when, and what risks are involved for the subjects (i.e. patients and users). The information we developed above, as part of the usability track and the information we will develop next on the risks track, will be very useful in putting such an application together. The information developed on these parallel tracks will be much more in-depth than probably required here for approval, but it should provide convincing answers and using one unified source should help keep all your ducks in a row.

A special reference to keep in mind while designing, conducting, and documenting clinical studies is the international standard on clinical investigation of medical devices for human subjects, ISO 14155. We discuss this in Section 5 and here it informs the clinical track of our metro map for good reason: it is recognized by much of the world as the golden standard for addressing how investigations are organized and conducted. Don't leave home without this stamp on your plans.

Finally, the importance of sound clinical evidence can not be overestimated. The allure of just checking a box can be strong, but the scrutiny placed on the evidence and its assessment later will be felt by everyone. A hastily prepared study may take as long and cost as much to run as a solid one, but perhaps only the latter will perform the function sought as enabling regulatory clearance. In other words, any mistakes in completed studies may be quite costly to correct via new studies and will push up timelines in painful ways.

Result data

Out of any clinical pilot study you'll get valuable data on the outcomes of the study and possibly also some measurement and performance data relevant to your device. This is particularly the case if your device has some measuring function or performs a calculation of some sort, as for example an algorithm working with physiological signals (e.g., ECG, EEG) might.

Depending on the nature of the study, the measurements may have been collected using some other device or a prototype of your own device. The dataset may or may not be available to you in full or it may have been anonymized or pseudonymized to some degree.

At the same, you may have undertaken extensive engineering projects to get your prototype up to the level where you can feel confident that any calculations it performs will stand up to established literature and the crucible of the real pilot study. These projects may have involved test runs, result comparisons, and result plots of different types that may very well all be of interest in assessing whether your product does what you think it does. The algorithms you now have at this early stage are not your final algorithms so we shouldn't call this work verification in the same IEC 62304 sense that we will be discussing later during the development phase, but as early proof that you are heading down a road worth taking and that this data will be very important.

In any case, you should now be accumulating valuable data on the performance of the device you want to build. You will, however, in all likelihood not have built that final device quite yet. Many things may change between now and when the ready device rolls off the production line, and the pool of guineapigs you measured may not match with the intended users and patients of the final device either. For these reasons we hesitate to call the data and experiences you accumulated here validation, but it may all be of interest to your ultimate verification and validation work.

The data accumulated here in the research phase is self-evidently important for convincing ourselves that the concept works, and for pulling the trigger on writing up any patent applications or scientific papers to be submitted (see discussion below). The data to be accumulated here will depend on the type of device concept you are working with and your judgment on what is needed to assess if the concept works or not. The standards and regulations rarely dictate any particular data modalities to be captured or even the exact nature of the data recorded, but instead it is up to you to make the assessment in the first place and then in the second defend it. There are, however, various guidance documents of evolving nature on the sort of questions you might be asking yourself during the development of a device, but aside from identifying those topics listed as stations on our roadmap these don't give any exact checklists to tick.

The above may sound like a cop-out, but what it really means is that in a world where medical devices range far and wide, and your particular brand of innovation might blow the lid off of anything that has gone before, you really are the best authority on what it is you are creating, what benefits it can bring, and how you can be certain that you know that thing is working correctly. Especially if you are raising your gaze up after years of conducting scientific research on some technology or physiological phenomenon, you already know what it takes to convince yourself of these merits. You thus know the kernel of evidence you will need to provide, after which you will only need to address the other topics on the roadmap, and be able to explain yourself to the regulatory authorities. A word of warning though: if you are looking to unleash the latest flavor of deep-space artificial intelligence out into the world and argue that the very nature of the technology is such that we shouldn't even try to explain it, or that it shouldn't be controlled lest we forgo some of the very benefits you are after, you may be in for a rude awakening. The medical-device world revolves around evidence and rationales, and not having data to back these up will make progressing on to the next phase very difficult, if not impossible.

Out of all the types of medical devices you might be accumulating data on through research studies, artificial intelligence is the one on everyone's lips today and the one swimming at the deepest end of the pool. On this front, the European notified bodies have put out a joint document ("Artificial intelligence in medical device – Questionnaire", Team-NB, 2024/11/14) listing the sort of questions you should ask yourself regarding data collection, data labelling, data pre-processing, and version control. Even if your device is not based on AI, these questions are worth considering. The questions include the following:

Data collection

- Have you specified where data is to be collected, and the number of records to collect along with a justification of why this sufficiently represents the target patient groups?
- Have you characterized the inclusion/exclusion criteria using relevant attributes?
- Do you have a defined process for exclusion?
- Have you described the collected data using descriptive statistics?
- Have you discussed the factors that could cause bias?
- Do you have a process to anonymize/pseudonymize data before use?
- How do ensure that patient data is adequately protected?
- Have you observed good clinical practice (e.g. ISO 14155)?

Data labeling

- Do you have a defined process for data labeling?
- Are your data labels derived from the intended use of your device?
- Have you justified the classification criteria?
- Do you systematically review the correctness of labels?
- Have you defined the competence of the personnel responsible for data labeling?
- Do you have a process for ensuring personnel competence?

Data preprocessing

- Do you have a process for preprocessing complete with description of the individual steps (including conversions, transformations, aggregations, normalization, calculation of features)? How is the correctness of each step checked and how does this relate to the risks associated with the step?
- How do you cope with different measurement methods and scales?
- How are values or metadata by the same name aligned?
- How do you handle missing values, outliers, and unusable data?
- How do you control changes to preprocessing steps?

Version control

- Are all the processes from above documented in sufficient detail?
- Is the process for data storage and retention appropriate?
- Are all software tools used also appropriately controlled?
- Are all training, validation, and test data appropriately controlled?
- Are all data sources (including clinics, devices) documented?

From the above summarized list, it is already apparent that the sort of questions expected from medical-device development are to a large extent similar to those expected from any serious scientific research project looking at measurement data. It should also be reiterated that the above list was developed for that most elusive medical technology of today, artificial intelligence, so if this list doesn't scare you, nothing much will. If scientific research comes as second nature to you, and you know your technology like your own pockets, you will know how to address data collection and the use of that data to prove the device you want to create either works or doesn't work. The overhead of medical-device development, if you want to call it an overhead, doesn't perhaps change those realities one bit, but it does introduce some other topics it wants to quiz you on: mainly those shown on the roadmap and on the list of questions above.

What you should pay attention to on the previous list of questions is how often a process is requested and a recorded justification is expected. Your ISO 13485 quality management system would take care of some of these (including software tool validation, competence management, documentation control), but if you don't yet have one to house your research activities you should note that these types of details are later required and you will at least want to record the who and the how quite carefully – and avoid trial and error approaches midway your project – so that the results can be adopted later. The data and activities you perform here will go a long way towards making the case that formal verification and validation should be undertaken, and show us what that might look like, but in most cases the work here will not replace that later work.

Patent applications

Academic institutions have gotten quite good at patenting over the recent decades, and perhaps some of them have always excelled at this. Years ago, I was amazed to hear that a certain southern university in the US, for example, makes significant sums from patents they hold for grass used on golf courses. In the past 25 years the drive to create patents and other intellectual property assets has grown at every academic institution we have been in touch with. Thus, it doesn't come as a shock that a research project aiming at a medical device should also look at patenting.

What may come as a surprise to researchers is that patent applications should be submitted before any results are published. Writing scientific publications will be on everyone's radar in an academic research project, but the same skills, drive and innovation that is used to shape papers should first be directed towards crafting patent applications around any key innovations we may hope to capitalize on later. Writing patent applications is an art and a science in itself, and where our hand feels shaky, we should again rely on experts in the form of patent attorneys to help steady it and avoid any structural flaws in the foundations we are attempting to pour.

A second surprise to researchers may be that it will often take a considerable length of time for patents to go through. In academia we are used to the review process of a paper taking one or two years with perhaps multiple rounds of comments from reviewers and editors needing to be addressed in the meantime. The same happens in patenting, only now each round of comments from the patent office appears even more hellbent on sinking the boat you crafted. You have a finite time to respond to these challenges which may be completely different from the previous round, and each round will cost you money. At the end of the torment, some 5–7 years have perhaps passed, and you finally get a patent – after which you start paying for the privilege of having a patent on an annual basis. In other words, don't expect patenting to be a singular event with a one-time payment.

The good news here is that even having a patent application will give you bragging rights long before you get a patent. That gauntlet your application has to run through is also a merit for your innovation: if it can't be thrown out right away you may have something in your hands that also business partners and investors might value. In many cases having an application is as good as having a patent for moving ahead in the early stages. At the same time, it should also be pointed out that patents eventually expire and will get marginally more expensive to maintain over the years. Thus, it's not always the best of ideas to accumulate patents in the drawer of your desk for use at some much later date. It may also make sense to keep something – the secret sauce or the Coca-Cola recipe – secret and only patent around that, but there we are probably getting into advanced fortunetelling territory.

Finally, one last thing to realize about patenting is that once you eventually get that patent it is not some magical shield protecting you from all unjust claims being hurled at your organization – it's really only a ticket, or a right, to be invited to defend your patent in correspondence with lawyers and in the court room, should things come to that. Not having a patent may mean your prospects in this battle are nonexistent, but having a patent may just be one part of the argument. All this again costs money, so it's a rather good thing we won't have to worry too much about all this at the start of our research project.

Scientific publishing

After working on patent applications, we will have made strides in solidifying our concept and our knowledge of what is truly new in the thing we want to usher into life as a medical device. Working on patents may have felt like an unnecessary sidetrack to those wanting to focus on science, but we would argue that it actually forced us to look at the most important question in any research project: the 'So what?' -question. Having now answered the question of how our innovation changes the status quo for the better we are also in a much better position to write an impactful scientific paper.

Often times the so-what of any research project is hidden in the conclusions section of a paper, hanging between the rows describing what was done and throwing around ideas on what could be done next. Since we had to really consider the topic in writing the best possible patent applications, we are also now in a great position to write our papers with more gusto and confidence. Pointing out the significance and possible impact of our research can be done without resorting to advertising talk and pasting trademark claims all over the place. It's not dirty to discuss the meaning of the work.

Writing scientific publications will be something well understood by most researchers, and something they feel at ease with. Papers published on reputable scientific forums will act as valuable proof of the value of innovation. These are artifacts that we can probably make use of later in explaining and promoting our device, but they can also be used to shore up our patents and perhaps create a buffer around the patents. After all, if we can't patent something that we accidentally already made public, the same will apply to others perhaps seeking patents around our innovation.

All that sneaky patent hardening aside, the best reason to write scientific publications around your innovation is to know your own pockets that much better and perhaps even elevate the level of scientific knowledge out there somehow. Many medical-device companies out there don't have anything innovative to write about, so if you do, and you write about it, that already elevates you above the pack in one very important respect.

10.2.3 Risks

Risk management is almost synonymous with medical-device development. Earlier in this book we looked at the many places risk management ties into our product development activities and how it can link up with usability engineering efforts. We thus need to start thinking about potential risks as soon as is possible when investigating some idea for a possible future medical device. At the same time, we don't know as yet, everything about the final shape of that device, how it will be used, and for exactly what purpose. If we don't know yet everything the device will do, we can't possibly know everything that could go wrong in using the device. This all makes it quite difficult to craft a complete view of risks and how they should be controlled.

We do, however, know the core idea we are working with, and this should give us some core concept of the final device. Not knowing everything else around that quite yet should thus help us cut through some of the noise and focus most on the most innovative – and hence probably also the most crucial and risky – part of the device. Our point here is really to turn a potential weakness into a strength: to go deep instead of trying to go wide, and be practical about both.

As argued above, we need some of this information on risks and our solutions for controlling the risks when we first apply for clinical study permissions. The information expected in the applications is probably mostly focused on data privacy issues and information security risks to study participants should vulnerabilities along these tracks be discovered. Thus, in ethics committee approvals we are actually looking at security risks more than the patient and user safety risks medical-device standards are mostly concerned with at the end of the day.

The segment of track for risks in the research zone only has a single station marked on the map, and thus the activities here can be slotted into a single group:

- Identified risks and their control

In the following we'll dive into this important first step of risk management.

Identified risks and their control

The number one thing to remember about risk management for medical devices is that it is always first and foremost about safety, i.e. avoiding harm to the patient or the user – and after that to the environment. Information security risks, too, are increasingly a part of the picture and should be addressed especially if they may harm the patient or the user. In general, risks can arise out of just about any part or aspect of a device, but perhaps especially out any changes in these or the context of the device. The driving goal in addressing all these is to reduce the safety risks around usability of the device.

The most common types of risks we should start to consider here include the following: safety risks, electrical safety risks, usability-related safety risks, and information security or cybersecurity risks. You might thus begin to think about risks along these lines and start to record the results of your deliberations in a spreadsheet of some sort. We might, however, suggest that during an early research phase you might be better off considering what risks might be involved in, for example, your concept, your technology, the underlying physiological phenomenon (if any), any possible comorbidities the patients may have, the characteristics of your users and patients, and the indications or intended use you might want to specify. Out of these deliberations you will start to build a valuable risk matrix that will be tremendously useful in later work but may also instruct the usability engineering work you undertake in this phase.

In constructing your risk matrix any format will do for now, but ultimately the format wanted is instructed by ISO 14971. The majority of the work in addressing risk management at the start will be in identifying and somehow prioritizing the risks, while the standard itself adds rigor to how those risks are classified, assessed, and controlled. This doesn't mean that you'd necessarily have to have a particular piece of software here as even a spreadsheet program would work, but some purpose-built tools may help you as long as they don't overwhelm you with features and workflows that take your eyes off the ball. For more discussion on risk management see our previous books Juuso (2022) and Juuso & Pöyhönen (2024).

10.2.4 What we should have at the end

As we wrap up the research phase, we hopefully have a concept in our hands that we feel is worth taking forward to becoming a commercial, life-changing medical device. We will have convincing early preclinical evidence on the utility of the device, and also data and reviews indicating that – also in an engineering sense – it all just works. We most definitely have formative usability engineering documents (including formative evaluation reports around any planned key features) that record the key ingredients of its recipe (i.e., users, use environments, and use) and will later speed up its development work. In other words, we now have a pile of documents answering to the correct terms, matching the development path expected of us, and all sorted neatly for later use. More importantly, we now know enough to know whether we want to move ahead or not. If we don't, we have the option to pull the plug now or somehow revise and expand our research activities.

We may have been running this research phase as an unpaid side hobby, as a limited proof-of-concept exercise, or a bigger focused research activity. In all these cases we have now dug around enough and asked all the right questions to know what it is we want to do and whether or not it could probably be done. Now comes the time when we ask ourselves whether it makes sense to do it. All sorts of black-belt business management wizardry and market seances can now take place and tell us whether our monopoly pieces can move ahead to the development phase, or if we should return to the starting square and perhaps tackle a different project altogether. We will not yet have considered everything over the product's lifecycle, of course, but thanks to the dosaging administered by the roadmap we won't have wasted time and resources scribbling all these premonitions down on the company letterhead either. There may even be parts of the product we are not quite sure how to build, but now we know whether we want to try or not. In other words, this is where architect Eero Saarinen hands over the drawings of the Gateway Arch in St. Louis, and the engineers start working out the details of how to realize the vision.

10.3 Development

The term "research & development" gets thrown around quite freely and in widely differing contexts. To some they are the same activity, but they may understand that activity as either developing a solution through research or as including an element of research in development. These two look the same and taste the same but are in effect playing two different ends of the same football field.

Companies have various approaches to placing that ampersand between the research activities and development activities. For them the line of demarcation may be based on budgeting, departmental involvement, or any number of practical considerations, but there is no doubt that the research should feed into development as smoothly as possible, or otherwise it will not be undertaken at all. For academia, research and development are probably intwined as concepts, but development in this sphere refers more to developing a solution for a research question than any concrete expectation of developing a product ready for launch.

In moving from research to development we may thus be changing the team out on the field. In drafting our roadmap, we wanted to pinpoint what activities should already be done during the research phase so that work there would proceed predictably and produce answers to the right questions. In moving to the development phase, we now want to be able to trust that these activities were done, and the related information is available in a documented form. We build on this previous work to extend our train tracks towards a final push on to the market. Our lines here are the same as we had in the previous phase with one obvious addition: product development. Let's dive in.

10.3.1 Product development

This is the play that all that rehearsing during research has been gearing towards. Here we go from our own 10-yard line to the endzone and hopefully go on to score a touchdown with our device, although the release itself will take place in the last of the big three phases we cover. In European football the lines will have different names, but the objective is the same: to sprint through the majority of the field and to score.

The activities can be roughly divided into the following groups:

- System requirements
- Software development plan
- Software requirements
- Architectural design
- Detailed design
- Implementation
- Test plan
- Test report
- Verification

We already cover all these activities in our previous book (Juuso & Pöyhönen 2024), so we won't try to reiterate all that detail here. The activities are based on the IEC 62304 standard, but as we point out in the earlier book, further requirements are placed on the development activities by the ISO 13485 standard on quality management systems, and these are just as weighty as the expectations of the software development lifecycle standard. In that book we provide a comprehensive mapping of the expectations, so please refer to that book for more discussion.

Instead, here we will discuss how the product development track fits into our wider metro map. You may notice that system requirements actually sit on the border between our designated research and development phases. This is because you may want to start developing a list of formal requirements based on your research already during the research phase. Doing so will not only give a jump start to your development activities here but may also mean that any early verification work you do in the research phase is that much more convincing since you have solid, documented requirements to verify against. The wealth of usability work you have already done may lead to stated usability requirements too, which you may have documented in the use specification, the UI specification, or even the formative evaluations for later use, but if you recorded requirements separately this too would be acceptable. It may also be debatable whether the requirements you previously specified are system requirements or fragmental software requirements, but this is the place to start sorting and revising those requirements.

On the map you will see links drawn between the software development plan and the UI evaluation plan. These two documents may evolve in tandem, and should match up so that the same user interface is built and evaluated. Similarly, new formative evaluations may take place as your detailed design and implementation mature. An agile development model in particular may give you partial prototypes to test, and this testing may be conveniently documented as formative evaluations even using your own staff.

At the end of the development track, you have a verified software build that constitutes a release candidate of the device you want to build. Here we exclusively discussed software development, but the same overall model applies to hardware development as well. For hardware, several other standards, including those of the IEC 60601-1 family of standards (see Juuso & Pöyhönen 2024), apply depending on the nature of the device you want to build. Also remember that if that hardware includes software the IEC 62304 standard we discuss here absolutely applies.

10.3.2 Usability

Here in the development phase of our activities, the usability track consists of the following major activities:

- UI evaluation plan
- More formative evaluations

- Instructions for use
- Summative evaluation

The usability track, thus, continues from the formative work conducted in the research phase towards the eventual summative evaluation. In preparation for that capstone activity, we develop the UI evaluation plan and the instructions for use (IFU). New formative evaluations may take place if they are called for by events taking place on the product development path. Finally, we conduct that most weighty of activities recognized by the usability engineering standard, is the one and only summative evaluation. Here it should also be remarked that the instructions for use are generally considered a part of the device and should thus be available for any summative assessment of the device.

The expectations for a summative assessment were already discussed at length earlier in this book (see Section 2.4.2), but as a quick recap the evidence we are looking for here is on the fact that our device can be used safely. Efficacy of the device is not necessarily a part of the story here (see the following section on clinical evidence for that), but safety definitely is.

10.3.3 Clinical evidence

Just as the usability engineering track moved for formative work via planning to a summative evaluation, here too we follow that same line of action. Here the activities on the roadmap are:

- Clinical evaluation plan (CEP)
- Ethics approvals for a pivotal study
- Pivotal study
- Clinical evaluation report (CER)

In unison with the above big-ticket items, we may also end up running new smaller pilot studies, revising the intended use, and writing up ethics applications, patent applications, and increasingly detailed new scientific papers. These are represented as faded grey stations on our map.

At the end of the day, the clinical evaluation plan and report are what is expected out of the clinical activities. These should provide the definitive assessment of the efficacy of your device – not just if the users liked using it, but if the device provided the value the intended use promises. The evidence may be based on literature and clinical investigations, but the verdict should be readable in the clinical evaluation report (CER). All the work on this segment of track culminates in the CER, and this is the second document after the intended use you will be pulling up anytime the clinical use of your device is questioned somehow. The CER is not a once-off document though, so it too will need to be revised over the coming years as instructed by your post-market clinical follow-up activities (see Figure 10.1.

10.3.4 Risks

Risk management is like the monolith from "2001: A Space Odyssey": no one knows how long ago it came along, but it should never go away. All the standards and many of the regulations remind us to point out that risk management should be a part of your activities from day one, and it should continue for as long as the associated product is on the market or in use, whichever takes place later.

In developing the metro map, we wanted to show how your appetite for risk management could grow from the early concepting phase to the ultimate release phase of a product. In this vein, we already began recording risks during the research phase where we argued that you had a singular opportunity to really focus on the core of your innovation. Here in development, we again get serious with our approach to risk management by first developing a plan for risk management. This plan should adhere to the ISO 14971 standard on risk management for medical devices. You may have heard of other standards for risk management from other fields of industry or being specifically tuned to artificial intelligence, for example. None of these other standards are acceptable as the spine of risk management in the development of medical devices, though. Such other standards may be helpful as something feeding into ISO 14971 risk management, but they cannot replace ISO 14971 if your device is to be a medical device.

After the plan is in place, and our risk matrix meets with the structure expected by ISO 14971, we are free to populate the matrix. We already know many of the important risks around our device from research and perhaps also know how we aim to control these risks. Here we build on our understanding of risks related to the concept, the technology, and any comorbidities associated with our patient populations and the indications we want to give our device. Failure mode effects analysis (FMEA) may be a useful link between our work here and that on the product development track (see above), just as usability risks are self-evidently the common language between our track and the usability track. Sooner or later, we will also want to address risks associated with information security or cybersecurity. Our designs will hopefully have kept an eye on these all along, but as things out in the field evolve every day the need to update our knowledge and assess the situation become greater towards the end of the development phase. At the end of the day, we expect to have a comprehensive risk matrix in place and a solid risk management report weighing the risks and the benefits associated with our device.

10.3.5 What we should have at the end

At the end of development, if this was the Manhattan project we might be happy with a huge crater, but in all other cases the thing we should now be staring at is the product we want to launch, minus perhaps some accompanying documentation and packaging we may still add. We should have a finished product, the evidence that it is what we wanted to create (verification), that it works (clinical evidence), that it can be safely used (summative evaluations), and a survey and action plan for any risks involved therein (risk report). With all these deliverables we can move ahead to the release phase, if that still is our intention.

Here we should thus ask ourselves the question if we really want to be chained to this product that we have now created. The medical device regulations place great expectations on us for overseeing and supporting the devices we release. This is not a catch-and-release situation, but rather a situation where we are expected to be planning for the college fund – and even a retirement plan – for the baby before a single ultrasound is taken. If we open the factory doors and ship out a medical device, we are to all intents and purposes tied to that device for the defined lifetime of the device all the way through to its planned decommissioning at the very end.

10.4 Release

Here we know that we want to bring a new medical device to this world and start putting the final package together. After running multiple tracks through the research and development process here all the lines converge, and we speed up to the actual release which culminates in the signing of the declaration of conformity and perhaps wrapping everything in cardboard and cellophane ready to ship out.

10.4.1 Product development

Moving over to the final straight before product release, the activities on the roadmap are:

- Validation
- Maintenance plan, post-market surveillance plan, and post-market clinical follow-up plan
- Accompanying documentation
- Declaration of conformity

We'll discuss each of these in turn in the following.

Validation

Occasionally it feels like the term 'validation' gets thrown around almost too frequently as a boardroom-credible alternative to saying we need to be sure of the answer to something before we can even begin to allow ourselves to ask the particular question. All this really means is that we know when we know, and until then we don't. In this it's perhaps similar to how 'gap analysis' is always the preferred term to throw up when we sort of know where we want to get to, but we have no idea where we are or how to proceed from there. We don't want to belittle either activity, they are both very important in the right places, but we want to point out that even if this is your first foray into medical devices you may already know much more about these than you think. This will particularly be the case if you have a practical background in the domain of work your invention relates to or in scientific research in the same field.

In the end, validation is one small word, but it carries a pretty big punch in medical devices. It is the gate you can only pass through once you know that the thing you built a) works as intended and b) also works for its user. The good news here is that you probably knew 50% of the answer before you even got started: you probably knew all along the sort of tests you wanted to put your invention through to know you could rely on it. The tests, trial runs, verifications, crosschecks, and other evidence you wanted for this – and perhaps already reported on in research papers – are all a very big part of clearing validation. If you have done that work to a high enough degree no one in medical devices will say that there's a special secret medical-device test that you need to run in order to prove your work. Let us rephrase that: everyone out there will always look to challenge your work and point at more validation steps to undertake, but if you truly know the technology and science you are working with you will already know all these and consider them conversation starters. Occasionally standards do point to tests that should be cleared, and these may seem arbitrary or disconnected from reality. A favorite anecdote of ours here is how an IEC 60601-1 standard requires a metal ball weighing 500 grams to be dropped on a device. A manufacturer may wonder in what real-world context this could happen, but the point is not that a ball is the projectile heading towards your device, but that the user wearing your device might ram into a door with the equivalent force. Tests such as these are not meant to brace us for spherical projectiles, but to standardize our benchmarks and provide real-world value.

On the other hand, as we already remarked above, if you still argue that your innovation needs to run free and wild out there in the world, and its special brand of cutting-edge artificial intelligence can't even be explained to us mortals, then you may have another thought coming. Medical devices as a domain are built on explainability, transparency and verifiable evidence so you must have convincing answers on these if you hope to launch your product. That brings us to the other side of the coin, that your invention works for the intended user. This was perhaps a tougher nut to crack during your hereto path, but leaning on the tracks we built for usability engineering and clinical evidence should mean that you also have the evidence you need on that front as well.

Given all the work done at the end of the development phase, here our task is to see that all the elements – verification, clinical evidence, summative evaluations, and a risk report – are available, speak the same language, and speak for greenlighting the approaching product release. The validation work might thus culminate in a validation report that doesn't regurgitate all this information but instead verifies that all these ducks are present and willing to move ahead. For more discussion of validation see our previous two books Juuso (2022) and Juuso & Pöyhönen (2024).

Maintenance plan, post-market surveillance plan, and post-market clinical follow-up plan

As we remarked earlier, the release of the device is not the end of our commitment. Instead, we must plan for the maintenance, post-market surveillance, and post-market clinical follow-up of our device. In other words, we must keep watch over how our device is fairing out there in the big world and whether its clinical context should somehow be re-assessed. We discuss these matters at length in our previous two books Juuso (2022) and Juuso & Pöyhönen (2024), so here we will gloss over the details of planning these activities. We will, however, dive into the topic of the post market in a moment in Section 10.5.

Accompanying documentation

This is where you ensure that the instruction for use (IFU) and any other planned accompanying documents are ready to ship with the device. You may also need to see to the translation of these documents if you had not already done so as part of the summative evaluations. For more discussion on the topic of accompanying documentation see Juuso & Pöyhönen (2024).

Declaration of conformity

The counterpart to the intended use statement is the declaration of conformity. The regulations see these two documents as offering fundamental insight into your device and as the bookends to your development work. After signing this document, you will ship copies of it with your device, present it to auditors over and over again, and effectively say that your device is now on the market. If there is a long space of time between signing the declaration and shipping the first devices you may want to make another distinction, but in a simple case the signing date is when you place the product on the market. This is also when your organization should have an ISO 13485 quality management system in place and be prepared for selling and supporting the device out there on the markets. For more discussion of the declaration of conformity please see our previous two books Juuso (2022) and Juuso & Pöyhönen (2024).

10.4.2 What we should have at the end

We have now built it, and they will hopefully come. We have a sellable product, and we have an organization backing it up with all the processes expected by ISO 13485 quality management and the various other standards our metro map fused into it. We have expenses and soon hopefully we'll have sales to offset those and perhaps try the upper boundaries of our bank accounts. In other words, this is where the egotistic manager in the cult comedy skit kicks the developers out with a Remington rifle and starts to make the "big bucks". In medical devices, however, this would be a shortsighted move as the "ever after" here involves much too much activity to just be faded out. That is what we'll discuss next.

10.5 Ever after

In the metro map, the bottom two rails, change management and post-market clinical follow-up, head back towards the starting square when all the tracks we have covered up till now have headed (at least most of the time) from left to right with a purpose. The purpose of these final two tracks of our model is to show how the longtail of medical-device development may connect with revising the tasks we have already gone through to develop a releasable product in the first place. These tracks show how we might address product updates and revisions to our clinical evidence. Let's discuss each track in turn.

10.5.1 Change management

At the beginning of our research project, some time back, the prospect of a successful product release seemed so distant that any notion of supporting that product on the market afterwards probably seemed about as remote as discovering extraterrestrial life outside of the whimsical domain of movies. Yet, that is exactly where we find ourselves now: having to carefully maintain and evolve the very thing we just carefully put together.

Thinking that we don't want to mess with something that works is a natural groove to fall into. Thinking that fixing this one small bug reported by a customer is no big thing and the fixed product will just be that much better after the operation is a good way to dig ourselves deeper into that groove. If we had a dollar for every time we came across a company that had burrowed itself into such predicaments, we'd still be poor, but well-grounded by all the ballast in our pant pockets. The lesson here is simple: something meticulously crafted will also have to be carefully maintained or it might as well have been put together just as haphazardly in the first place. This does not mean that we have to redo everything with every single update, only that we must consider redoing everything and revise the parts that need to be redone. This is what the metro map tries to visualize in showing us four access tracks to re-enter our development path.

We discussed the A.B.C.D model of configuration management in detail in our earlier book, Juuso & Pöyhönen (2024). Each level of change is represented by a particular access track in the metro map. The tracks are:

- **D-level change**
  The smallest change might be a security patch, a minor bug fix, a minor update to a third-party software library, or perhaps something only affecting some embedded accompanying documentation (e.g., the instructions for use). Thus, to make the change we would reconsider the continuing validity of our development phase information (risks, verification, usability and clinical evidence, etc.) and proceed accordingly. The change could thus be enacted very quickly without retouching the vast majority of our earlier work.
- **C-level change**
  If we had a bigger bugfix to perform, such as a previously missed error in some feature or a minor tweak to some existing feature, we might still be working against the same planning documentation and move to the implementation step. The work should still be documented in detail even if we argue that we are just following those earlier plans even more carefully. The activity might also spawn some corrective and preventive action processes in our QMS that might outlive the bug correction process itself, but the correction could still be enacted fairly quickly.
- **B-level change**
  In case of a more significant fix or a new feature to be added which might affect risk management and clinical evaluation, for example, we could return to the requirements analysis step and work forward from there.

- **A-level change**
  The most significant level of change might take us all the way back to step one and have us reconsider everything from our intended use onward. This would probably lead to reconsidering all our work, but again not necessarily to redoing it all.

Note that in all the above cases the final leg of the tracks from the risk report to the declaration of conformity will need to be addressed in its entirety. After making any change we should reconsider the status of our risk management, product realization, usability engineering, and clinical evaluation information. Only when this key set of information is valid and supportive of going ahead should we move ahead.

The view the metro map gives of change management tying back into our development lifecycle is something we have not seen anywhere else. We feel that this way of visualizing the connections is an insightful tool to help us consider the types of changes we might actually have, how these should be addressed by our development organization, and how these paths could be optimized. You might decide on a different number of magnitudes of change in your configuration management scheme, or you might choose to connect the wires in some slightly different ways. Regardless, we believe the novel visualization here will assist you in analyzing your processes – and also, should you have been so inclined, sway you away from thinking that a quick fix should be free from any processes whatsoever.

If the above sounded too easy, you might also consider how the metro map would look when multiple change orders are running on the same tracks at the same time. This is a situation we touched on in Section 1.23.

10.5.2 Clinical follow-up

Similar to maintaining the device via controlled updates, also the clinical evidence for your device might need to be updated. To this end, we have included a special post-market clinical follow-up track on our metro map. The track here is developed, first of all, to remind you of the need to maintain clinical evidence once a product is released, but secondly to show how the periodic safety update report, a review of the post-market clinical follow-up status, and a new clinical study might be positioned in relation to all your development activities.

The idea here is that a new study would take us back to reconsidering our intended use and revising the activities on our clinical evidence path from literature surveys to pilots and a revised clinical evaluation report. Any new discoveries might, of course, affect our activities on the parallel tracks. Unlike perhaps in the case of the activities on the other tracks, an updated CER might be the end of our run here as no changes to the product would be required. The next time we would be making a change to the product, we would perhaps have a new revised CER waiting for us at this station – or even just the verdict that a new PMCF study was done but it had no effect on the CER. After a new release is eventually made, or if a certain amount of time has passed or some other trigger is reached, a new PMCF cycle could be run with similar effects.

Similarly, a periodic safety update report and a PMCF status check might be all that is needed during a cycle through the shortest loop via the D track. In this case, too, we would review the other key information from risks to usability and proceed as warranted.

10.6 Considerations for a startup

The metro map was borne out of our very real crucible of developing a quality management system (see Juuso (2022)), a software development pipeline (see Juuso & Pöyhönen (2024)), a medical device, and a medical-device startup all at the same time. For this reason, we had to think long and hard on what activities to tackle when and how so that we'd make it to the next trench of some milestone-based funding promised to us. It also meant that while we were considering the activities represented by the stations on our map, we were also quite busy seeing to many other more business-oriented activities and deliverables. These are plotted on the sidelines of the map in grey print.

The small activities at the sidelines of the map, and in particular below it, are included here to provide helpful pointers to activities that might be topical at each stage if you are also trying to launch a startup to release the product you are developing. In a startup everything needs to be done at the same time, and you will often need to be the one doing all of that, but we feel the progression shown by these side items is at least anecdotally helpful. Note that the items connect to our principal left-to-right path from research to release, not so much the change management and PMCF paths.

In embarking on the train journey shown on the map, a startup will first want to consider whether their device is a medical device or not. The path shown here is only expected of a medical device, but it is also beneficial to a wellness device, particularly if that wellness device might one day become a medical device. Secondly, the startup should consider what is the minimum viable product (MVP) – or the most concise concept yielding the maximum impact as measured by novelty and the impact only you can make. We might call this shaping the concept, or cutting the steak in the way you really want to. Aside from Steve Jobs getting obsessively involved in the plastic box for the computer he was working on at Next while briefly exiled from Apple, we don't know of many people whose innovation is about the plastic box or the cardboard packaging around the device they actually want to usher into being. Don't get distracted by the things you don't need to do, is our message here. Focus will help you achieve more where it matters.

Related to the above considerations, you may also come to the conclusion that your innovation would be happiest residing inside someone else's product. This might particularly be the case if the device you wanted to create was a software algorithm. In such a case you might realize that not many of the medical regulations apply directly to you but instead apply to the businesses who will be your customers. In such cases, of course, the regulatory requirements might apply to you because your B2B customers say so, or because you need to have convincing answers on these requirements to attract new customers. In any case, the roadmap we have presented above will have you aware of the whole sphere of expectations and producing the right deliverables for your phase of development.

Finally, as we are discussing startups, attracting new trenches and rounds of funding is never far from the conversation. The same clarity of the road ahead and certainty of knowing your position along that road the metro map brings to development activities, it also brings to your conversations with investors. Knowing the big picture and being able to plot your place in it will be part of every single conversation with investors, and the roadmap will help you address the parts under your control. The whole of the conversation in this section was aimed at illuminating the most practical and to-the-point path through medical-device development, so we hope that particularly startups will take note and benefit from the thinking here.

10.7 Final thoughts

If we try to focus on everything at the same time, we are actually not focusing on anything at all. If we try to skip to the finish line, we will probably have to backtrack and do some required activities we neglected in between. The metro map we presented here served two main purposes. Firstly, it showed us the path forward so that we wouldn't need to wonder where to start or to keep guessing on what we might have to do next. Secondly, it allowed us to focus our resources and acumen on the immediate job at hand while aiming systematically at a final device. This it did by showing us the topics we should focus on at each stage – focus on to move to the next level, but also focus on because we actually were well-versed to consider those topics there and then. The map allowed us to pace ourselves in a thought-out way without overwhelming us with everything that needed to be done all at once.

One of the American founding fathers, John Adams, famously wrote in a letter to his wife in 1780 (the language here is modernized):

> I must study politics and war, so that our sons may have the liberty to study mathematics and philosophy. Our sons then ought to study mathematics and philosophy, geography, natural history and naval architecture, navigation, commerce and agriculture in order to give their children a right to study painting, poetry, music, architecture, statuary, tapestry and porcelain.

This measured, thoughtful, and carefully planned-out doctrine is a grandiose way of saying that we hope our roadmap proves itself useful for researchers wanting to become product developers, developers wanting to become manufacturers, and manufacturers wanting to keep churning out useful medical devices to their users. It is a cliché to say that the only thing worse than the wrong decision is indecision, and while that doesn't readily apply to medical devices where lives may be at stake, we hope the map we have drawn out here allows us all to plot ahead, focus on what matters, and move ahead as appropriate.

Was that it, you may ask? The metro map we presented here may feel like it doesn't tell you enough about the details of each activity. It may feel like the map doesn't give you the one checklist you want before releasing a product into the wild. Both are fair criticisms, but they also miss the fundamental point of a roadmap: a map should focus on giving you the lay of the land and point out the stops of interest on the way to the destination you want to get to. It shouldn't try to teach you to drive a car or fly a plane at the same time.

In discussing our model with academic researchers, we have occasionally heard the criticism that the map should be presented as a step-by-step guide with checklists to take our railbike to the next station along the track it happens to be on. There is certainly a need for station-specific checklists and guidance, and we have done our best in the past sections and our two previous books to provide that practical help on each activity. Ultimately, we feel that to present all that advice in a single graph would be as feasible as it would be helpful – which is to say neither, not actually.

The constellation we have teased out here will give you a good idea of the network of activities expected of you on the path to releasing a medical device in the real world. It gave names to the common activities and placed them in a more-or-less chronological order from concept to product via research and development activities. The most paralyzing uncertainty out there is not knowing where to start, then followed by not knowing where to go from there. Hopefully the metro map we have presented here helped you answer both questions – it certainly put the 'start here' sticker on the map to developing medical devices.

Now that we have mastered the development and maintenance lifecycle, we can begin to look into the future and consider some of the recent and emerging challenges to usability engineering.

Chapter 11

Seeing into the Future

In this light-hearted section we peer into our crystal ball and point out some recent trends and current topics that may impact how we all approach usability engineering in the near future. Generative artificial intelligence, as depicted in Figure 11.1, *is necessarily a part of our discussion here, but so are cybersecurity, remote access, and the cautious proliferation of medical technologies, to name but a few of the topics we'll briefly discuss next.*

The former German Chancellor, Angela Merkel, famously remarked some years ago that it may not be too long before humans need a special kind of a driver's license if they want to drive their cars instead of the onboard computers. Another thought leader showed a video from a street intersection somewhere in South Asia as an example of traffic flowing simultaneously in all directions without any traffic lights being used. Both examples illustrate why a mixture of human intelligence and artificial intelligence working at the same time may have its issues. In many presentations of autonomous driving vehicles human error – or just general human unpredictability – has been the berated cause of many kinks in an otherwise perfect, programmable world.

Until humans are completely out of the loop – which won't happen in our lifetime – there will always be a need to address usability of the devices we build for ourselves. This is not to say that usability engineering in the 2050s will be like usability engineering in the 1950s as the world around us, the expectations of what is possible and what might be possible or even probable have changed all through those years. We must, however, consider the pros and cons involved with each new technological innovation

DOI: 10.4324/9781003645702-11

Figure 11.1 AI is asked to grant more wishes than just the three asked of Aladdin.

and how these may affect our perception of devices and our interaction with them.

In this section, we'll discuss some of the big shifts and current undercurrents impacting the world of medical devices and usability engineering. Some of the discussion is centered around possible challenges while the rest focuses on possibilities brought by each topic. Most come with both – with great power comes great responsibility, as Spider-Man would say.

11.1 Software

Software used to be a strange new topic, at least in the last millennium, as Figure 11.2 depicts. Today software is so ubiquitous and its applications so diverse that discussing software may feel as ordinary as discussing nuclear power flowing from the wall sockets or using RNA technology to combat viruses via inoculations. What hasn't been said of software by us in our previous two books or by others over the past 50 years? Perhaps not much.

In terms of usability engineering around software, perhaps the most important thing to realize is that software does something under the hood at the behest of the user. It may be simple enough so that we as users feel that "what we see is what we get", and all the behind-the-scenes work can more or less be taken for granted. When we type a key on the keyboard, we expect to see a letter shown on the screen in front of us. When we

Figure 11.2 The mythical, magical, and the mundane software disk.

type in a calendar entry on the phone we expect to be reminded of that event on the date and time we specified. When we ask Google to find all the pages about a vacation destination we plan to visit we probably won't sue the company when some sightseeing spot gets omitted from the list of results. By the same token, a reflex hammer used by a doctor to produce a knee-jerk reaction is quite straightforward to use and the results easy enough to interpret, but a piece of software can look at something quite a bit more complex, like the presence of some recessive genes in our DNA.

Software today gets used for a massively varied set of tasks, and it would be futile to characterize them all at the same time. The expectations, inputs, and outputs vary greatly, but what is constant is that something we want done gets done without us doing all the manual work ourselves. Perhaps the ultimate way to think of software in terms of usability is to think of it as outsourced value adding that is done as a result of a prompt or an input from us, and this results in some output being handed back to us. The controls available to use in specifying that input and the concepts to be understood by us as users in giving that input or understanding the resulting output will vary, and therein lies the difference between good usability and bad.

The second key characteristic for software is that the machine under the hood will change. Since we aren't forced to come face to face with the inner workings of that machine everyday, we may not even notice that the machine has been updated somehow. When Google changes the algorithm behind its search engine, the first people to notice are probably the search engine optimization (SEO) folks whose bread and butter it is to ensure their clients' websites show up at the top of the search result lists. We as casual consumers of the web are less likely to notice even if that realignment of the information we see online could be redirected in some significant, but incremental way.

The above issue is much more pronounced when a software update changes the way a medical device works. If a new alarms threshold now causes an alarm to occur much more frequently than before, the nurses using a remote monitoring device will probably notice the difference. If an electronic information system starts dropping letters from the end of entries that are too long for the database, we may not notice anything for a long time. The beauty and danger of software is that it saves us from fiddling with the nuts and the bolts of everything on a manual, grass-roots level and instead lets us focus on the task in front of us on some higher more meaningful level. This abstraction or value-add is welcome, but it also means that bugs and ants of many sorts may reign over the lower abstraction levels.

None of the above is anything new, but it means that in terms of usability we are faced with more issues than we were when tree bark with medicinal value and sufficiently sharp knives were the pinnacle of medical devices. Any time we move something the user used to take care of themselves to underneath the hood we should consider how to create practical transparency through that hood, or at least ensure the concepts and controls cut through the hood make sense to the user.

My father is an avid fan of the Swedish-made SAAB cars. The reason he loved the now-defunct car maker was that the cars allowed him to read the road and stay in touch with the road conditions and the handling of the car much better than some other brands out there. He didn't have to operate a manual choke to control the engine, or wind a crank to get the car going as Henry Ford might have done. There was already wizardry under the hood in the form of circuit boards and software, but not as much as what exists today. The hallmark of good usability was there though: my father liked what the car was telling him and how it handled when he applied the controls given to him. The same is an important factor for all human-computer interactions today: we either understand what the computer does for us, or we don't, and this will largely determine whether we use the software or not. In reality it's of course not a black-and-white thing but a sliding scale of some sorts.

The abstraction afforded by software is and will be the key challenge to solve in usability engineering activities for software. The benefit of doing something faster and saving cents and dollars while doing so is, for our purposes in this book, a collateral effect at most. Even if the software does something we couldn't do – like do a Fourier transform for a massive dataset or troll through an ocean of internet data – its output still needs to make sense to us in concepts familiar to us. Ensuring this interface between the human user and the computer running software is the point of usability engineering – and much more so than just competing with other software out there on whose interface looks the prettiest. The factors to consider in this are diverse enough and changes in available building parts, libraries, and design philosophies volatile enough to keep us perpetually occupied, but in terms of usability for medical software we know enough about it all today to plan comfortably for the future.

This entire book has discussed usability for software so there is not too much we need to add in the conversation in this section. One thing we would add, though, is that software used to be loaded with expectations of omnipotence, or software running amok to take over the world or at least break out from under the hood somehow. Today most of those fears have shifted to artificial intelligence, which we will discuss next.

11.2 Artificial intelligence

We load artificial intelligence (AI) with expectations of doing the things we can't do and the things we can't be bothered to do. At the same time, we fret that it may evolve into something that hurts, kills, or just puts us aside after gaining some level of consciousness. In short, we are probably liberally writing cheques left and right, both good and bad, that AI can't cash for a long time to come. AI is, in many ways, the modern shorthand for the unknowable next generation of software, as Figure 11.3 hints. It may be the Jedi Force telling us everything is okay or the Skynet of the Terminator films out to kill us all. AI, however, is only ever a one single thing in films – in life it is everything from machine learning to magic and it is used for an increasingly vast range of applications in the real world. Our fears, dreams, and expectations in each of these slivers of AI application vary too. It is the topic of software engineering, usability engineering, risk management, and careful validation to ensure the interface between (wo) man and machine works smoothly.

Figure 11.3 AI is the new incarnation of the placeholder for our hopes and dreams.

Everyone and their grandmother will remember to tout the dangers of artificial intelligence, how it will run away with both our jobs and our understanding of what is happening under the hood. With AI taking over part of the work and doing some value-add processing in between it is necessarily the case that our medical devices are no longer quite as WYSIWYG – what you see is what you get – as the knives and scalpels of yesteryear.

This may happen, but also the opposite may happen. The use of AI may mean that the handles we work with are also more intuitive or harder to use erroneously than what was previously best practice. We could be dealing with detected areas of interest in some radiological images, or with automatically extracted counts of cell clusters in some tissue sample, and this could be easier to comprehend than the physical sample glasses under our microscope. The controls, too, may be more intuitive than having to adjust some voltages or concentrations of liquid stains. AI has the potential to change all the buttons, dials, pulleys and levers we might use in operating medical devices to some more user-friendly concepts – move the envelope from the engineering domain to the domain of medicine in discernible increments – but it may also turn all the labels into indecipherable gibberish for the intended user. There is, however, little doubt that artificial intelligence is also of interest in the domain of usability engineering.

If after the introduction of AI everything works as before, nothing may have changed in terms of usability. It will still be necessary to validate the software and ensure our risk management is up to its job, but if usability has not changed it hasn't changed. If, however, we come to trust the AI having done everything and having considered everything we may be setting ourselves for a nasty surprise when it hasn't caught some exception or skews our attention as medical professionals going over some vast data sets to only look at those entries it was drawn to. Things happen under the hood, those things evolve, and all this means it is necessary for us to understand at least on some level what is and what is not taking place under the hood. The greater the magic under the hood the more we are asking from the user in terms of faith and less in terms of oversight. The more the user is taken out of the machinations the more loss of context they will experience and the less command they will have over the process and the facets of the information involved.

At a recent event focusing on AI and human rights, a senior researcher in the medical side of the field pointed out that doctors have around 15 minutes of time to spend with each patient. The interview, diagnosis, and reporting all take away from this time allowance which does not leave much time for also explaining the inner workings of AI to the patient. To make things more challenging, the patient may already have asked generative AI to diagnose them and may have a solidly hallucinated diagnosis to confirm or to dismantle. Added to this, the doctors of today may not always be in a position to fully understand and thus also explain AI to the patient. The point of the research was that we perhaps can't yet ask doctors to explain AI to patients, but we need to find ways of helping them do this, and we need to get better at this over the coming years. How we make AI transparent and explainable is a key concern when addressing all the stakeholders involved in this equation: the patient, their relatives, the doctors, the nurses, the manufacturers and whatever parties may be relevant in our context.

Work on AI and even the application of usability engineering to AI is underfoot (keep an eye open for IEC 62366-3) and much more thinking will be published on good usability engineering for AI. Similarly, work is proceeding on the application of medical-device risk management to AI/ML (look for ISO 24971-2) and a great many topics from the management of AI products to addressing bias and the ethics of AI, to name but a few. For the next decade AI will quite likely be the major proving ground for most of human innovations. Yet, it may help to ground our hopes and our fears to remember that while the term may only be 70-or-so years old, the concept itself has surely been around since at least antiquity. AI is just the latest kind of fire humanity has learned to make in the last 100,000 years, and just like fire, we need to harness it appropriately. To put that another way: there is no FDA product code or EU MDR code for magic, but devices based on other more explainable flavors of AI are already routinely entering the market.

11.3 AI talking to AI

This may sound like sci-fi, but what if it wasn't us interacting with the AI system? What if some instance of AI was conversing with another instance of AI, as illustrated in Figure 11.4, and we as the doctors, nurses, and patients would be none the wiser until the two AIs agreed that we needed to be told something. At a dinner held for AI researchers a few years ago, one professor joked that even AI had to talk with his wife before taking over some aspect of his life, but we felt this was missing the point: his AI and her AI will probably stealthily conspire to take over both of their lives sooner or later. The question is how well does the AI know us and what is in our best interest.

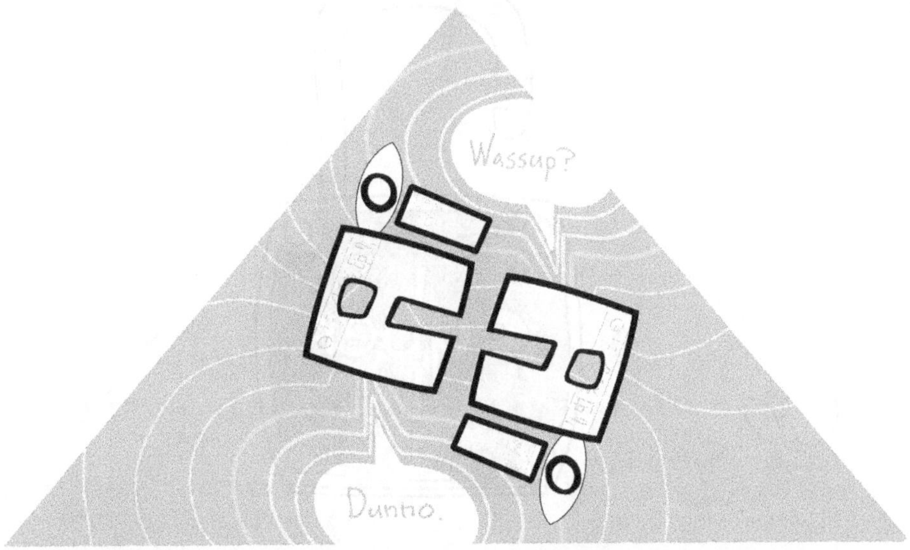

Figure 11.4 What do AI's talk about when they are alone?

How will user interfaces have evolved when the interface appears to be more of a one-way street: the AI decides for us and only informs us of the outcome instead of having us twist knobs and punch keys. Is this still an interface, or is it something else? For now, this is science fiction, but for us the point is that AI has the potential to change the inputs and the outputs considerably over the coming decades. Usability engineering, too, will need to stay alert to stay relevant. Maybe the question is not whether androids dream of electric sheep, but how do android dreams affect us sheep? Stay tuned to find out more.

11.4 Cybersecurity

Usability and cybersecurity both feed into the risk management of medical devices, but that is not the extent of their common ground. Cybersecurity is part of the fabric around medical devices, as Figure 11.5 suggests. A user no doubt hopes the device is safe without any effort from them, and much pressure is put on the manufacturers of offending devices or the parties responsible for providing cybersecurity-related updates to devices. Yet, a safe device may be deemed too difficult during installation and

Figure 11.5 Usability engineering may help unlock cybersecurity issues.

vulnerabilities appeared left and right by the user for the sake of just getting something to work for them. A typical example here might be opening up a buffet table of ports when attempting to get some piece of software to work.

The reality is that safety updates are necessary to keep the walls of our fortress standing, but if applying the necessary mortar and bubblegum becomes too laborious or incomprehensible it will more easily be forgotten. If it's too taxing to stay current with updates some users will want to bury their heads in the nearest sandbox. Many professional users will also take refuge in the knowledge that the hospital network will protect them from outside threats, which is true, but not necessarily foolproof as time goes on.

Standards such as ISO 27001, IEC 81001-5-1, and IEC 82304-1 will try to help, but they don't organically involve usability engineering. Guidance from the FDA, MDCG, and the European TEAM-NB organization will also add their voice into the mix. Usability engineering is always included in this conversation via risk management and occasionally even given a big voice, but the one realization we feel is too often lost in the cacophony of good intentions is that the usability of maintaining cybersecurity also matters greatly. It is not enough to assert that any emerging usability issues are addressed via risk management when we should ensure that maintaining cybersecurity from the user's point of view should also be made as simple, straightforward, and rewarding as possible. Making installation as complicated as defusing a nuclear bomb and expecting all users to know better is a recipe for at least something to go wrong some of the time. Asking the user to always log in using a separate account to ping for updates or install them is nuts. Pushing out longwinded notices that require specialist skills to interpret does no good for most users. Running automatic updates that interrupt the use of a device are perhaps a necessary evil, but occasionally the harm may be greater than just annoyance.

If only a device could somehow tell us that it was going bad like a banana, or if our friendly background AI could tell us that it has noticed something a little off it wants to act on, we might be on to a solution for cybersafe devices. Until then usability can help us streamline cybersecurity in ways that may even address the root issues behind many reoccurring vulnerabilities. A concrete first step here is to provide practical installation instructions, a clear cybersecurity status indication, and a convenient way to stay current with updates. Both then and now the aim should be to let the user know what is expected of them, why this is expected, and how it can be achieved with as little friction as possible. This is what usability engineering can offer to cybersecurity. Otherwise – whether you're a manufacturer or a user – expecting cybersecurity to always be handled by someone else somewhere will be akin to an endless game of whack-a-mole: many threats will be thwarted, but you can bet that some threat will be realized sooner or later.

11.5 Remote care

Earlier in this book we talked about using simple medical devices that are more or less 'WYSIWYG', what you see is what you get, in nature. We have also talked about how AI can sometimes make it harder but occasionally also more intuitive to see what a device is doing. Another contextual change we see affecting the usability of medical devices is the remote use of those devices, such as setting up centralized control rooms to oversee patients at a hospital or across several different care facilities. In these environments, 'WYSIWYG' may be more about you see is what you get, rather than about seeing a patient in front of you at all times.

Our intention is not to downplay the value of remote care in providing quality healthcare. We see remote care and remote monitoring as both necessary and welcome technological solutions to enabling access to world-class care across geography and time, as illustrated in part in Figure 11.6. In terms of usability, the context is markedly different than a live setting, though. The difference may stem from not seeing the patient around the remote measurements, or it may come from not being physically present in the same space (e.g., not hearing the other beeping devices in the room). Occasionally it may also happen that accidentally pressing a switch in the remote-control room has an effect in the room that would be hard to miss if present, but may be missed as a small label on a screen (e.g., switching off the lights or AC in a room). If the rooms monitored are located across multiple time zones or different climates the differences may grow larger still. The same may also apply to the use of in-vitro diagnostic devices where the connection with the patient may also be remote, but at least this connection is usually moderated by a medical professional calling for the exams and interpreting their results to the patient.

None of the above represent issues that couldn't be solved. They, however, need to be identified to be solved and this is where usability

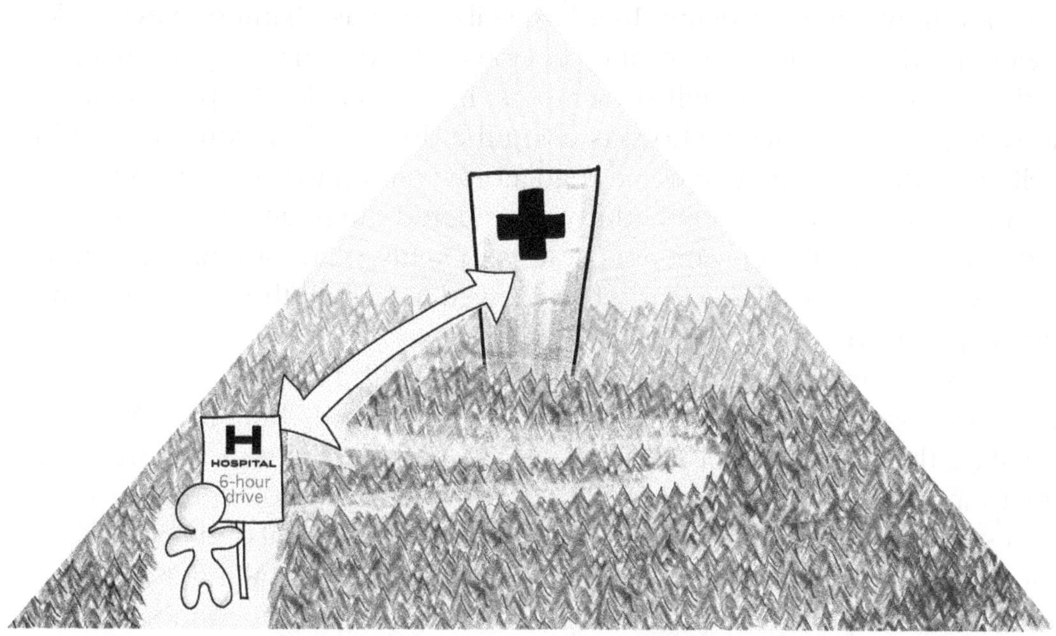

Figure 11.6 Distance has its advantages and disadvantages for seeing clearly.

engineering may be leveraged. Seeing the patient from a distance should facilitate access to all the pertinent information and also make it as hard to miss important additional factors that occur from being with the patient in person. Reacting to an event may require actions at both ends of the remote connection: acknowledging an alarm remotely is self-evidently not the same as holding the arm of a patient to make sure they are alright. A surgical robot could be controlled from afar, vital signs can be monitored across the internet, and many other things will be possible regardless of space and time, but being present and having access to a virtual environment connected to a real one will be two different things from the perspective of usability. The nature and degree of the differences will vary from one application to the next, but it is our responsibility to not take it for granted that a condensed, virtual view always provides the whole truth of the matter.

An anecdote we can offer here is the standard map view of a ride-sharing app. Once you order a ride, some apps show the driver on their way to finish a previous ride and dropping off that passenger before coming to pick you up. This information may be practical to figuring out how long it will take the driver to arrive at your location and from which direction they will be arriving. In a large city, such as Miami or New York, seeing the drop-off location of the previous rider doesn't tell you much as the buildings often are tall skyscrapers. In a small city the pin might, however, tell you exactly who was using the driver before you. This could well be a privacy issue that some regulators would raise an eyebrow to even if the eager developers had not considered the issue. Our message with the tangentially relevant anecdote is to softly point out that sometimes the obvious becomes less clear when viewed across all the space and time our scope of operations may involve.

The previous story had precious little to do with healthcare – unless you attempt to use GPS to locate patients inside multistorey building. Our goal in telling the story was not to deter you from embracing remote care as an environment. Remote technologies help provide access to expertise over both space and time, and make healthcare more universal and socially just, but it has many new usability characteristics to address. Taking a device that was used locally and enabling its use remotely is about much more than just flicking an internet connection on.

11.6 Preventive care

Taking the previously discussed notion of remote care one step further we can consider the effect preventive care might have on the usability of medical devices. In many ways this changes nothing: a feature we look for was either found or it wasn't, a value threshold was either reached, or it wasn't, and all of this either says we think we have an issue, or we don't think we do. Occasionally the result may sit on a fence, as Figure 11.7 attempts to illustrate. Professional users will, in any case, use the device and interpret its message without always seeing a real patient with visible symptoms in front of them. A lay person might use the same device and be inclined to hypochondria or use any excuse to brush the worries aside. The device we are discussing might not even be a medical device as it only works across some population-level data and gives us vague overall ponderings on the future prevalence of this or that disease.

At the same time, we could be talking about an AI system making life-and-death decisions on a person's life – or at least making the call between dismissing any concerns or setting out on a path of small actions that could thwart off a major future ailment. For now, the users of such tools may be expected to have a high degree of specialist skills in interpreting

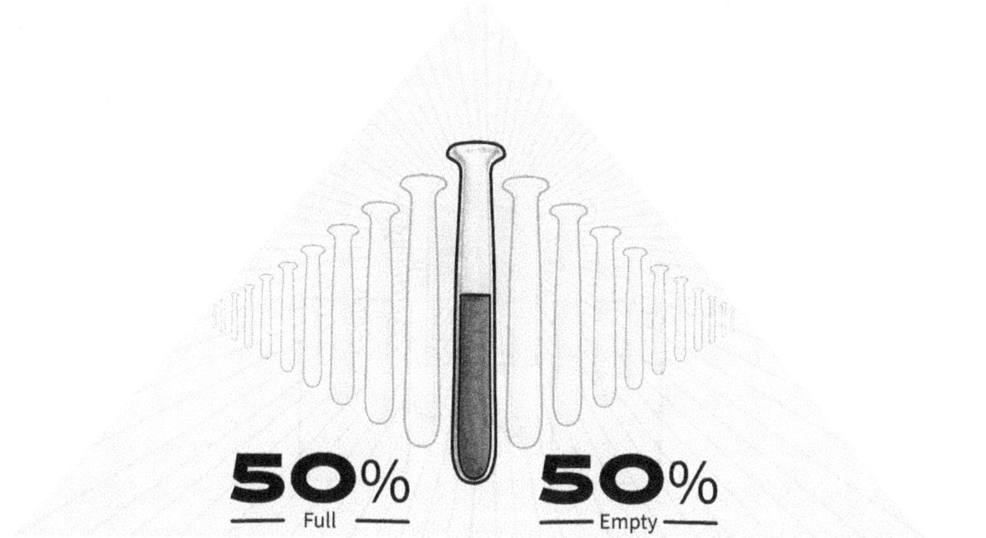

Figure 11.7 How many nanomoles is the difference between something that is 50% empty versus 50% full?

the statistics and medicine involved, but this may not always be the case going forward. The question therefore might be how does the usability of a device change if we only see the impact years from now? How do usability characteristics change if the patient themselves uses the device repeatedly over a span of several years before seeing any effect, if even then? How is usability affected if the user never sees a patient at all? The questions can be derived from usability activities today, but whether you should dive into them or not will depend on the type of device.

11.7 General purpose technology

As technology evolves it's not just medical devices that get more sophisticated and ubiquitous. As playfully depicted in Figure 11.8, we have already seen anecdotal examples of general-purpose internet of things (IoT) devices overwhelming hospital environments where similar medical IoT devices have struggled to get their voices heard when the channels have been blasted full by ill-behaving devices not approved for medical use. Think of a simple light switch that seems innocuous enough but then blocks the frequencies by relentlessly pinging out radio messages – whether this was by design or by accident the result may be a denial-of-service type of attack on the environment.

Figure 11.8 The one that does not fit in?

With the previous in mind, it may not be fearmongering to remember that much of the development in artificial intelligence takes place outside the strictly regulated domain of medical devices, and sometimes such AI-enabled devices too may come into contact with medical devices. So far the only stories of non-medical AI encroaching on the medical domain have been via what jokingly could be referred to as man-in-the-middle attacks: via generative AI convincing a patient who then quizzes or questions their doctor on whatever the AI happened to say. In the future, medical AI and general-purpose AI will increasingly interact, especially as the latter is likely to dominate wellness applications outside the stricter medical domain. If we are not careful, this mix up of AIs may end up being the equivalent of the industrial silicone case that brought about the stricter medical-device regulation in the EU after a series of adverse events were reported when industrial silicone was used in breast implants instead of the medically approved kind.

Yet, there is no reason to panic or overreact here. Both medical technology and non-medical technology will evolve and will increasingly interact with each other. In some cases, usability engineering may help to negotiate this interface and in others it's more about cybersecurity, risk management, and design controls. We bring up this topic here as a future challenge and expansion to the environment our medical devices are intended to operate in. The mixture of medical and non-medical technology in an environment is not, for now, explicitly addressed by the IEC 62366-1 standard, but the tools provided by the standard are, of course, applicable.

11.8 Proliferation of medical technology

Hand in hand with the previous section, also medical technology is breaking out of the once strict confines of the doctor's office, as illustrated in Figure 11.9. We are still a few decades off from everyone having an MRI or an X-Ray machine at home, but other forms of technologies are venturing into the domains of home use and even prescription-free use. Even something as profound as neural stimulation is now, in its non-medical and perhaps non-scientific branch of neural hacking, partially the domain of enthusiasts. These enthusiasts may be equally knowledgeable in their field as some doctors, or they may quite worryingly be driven by social media posts.

The limitations in education, skills, and even attention spans of such users may differ and somehow the device manufacturers should still ensure a level of safety over all interactions with their devices. Achieving practical transparency of information while preventing jumping to easy conclusions

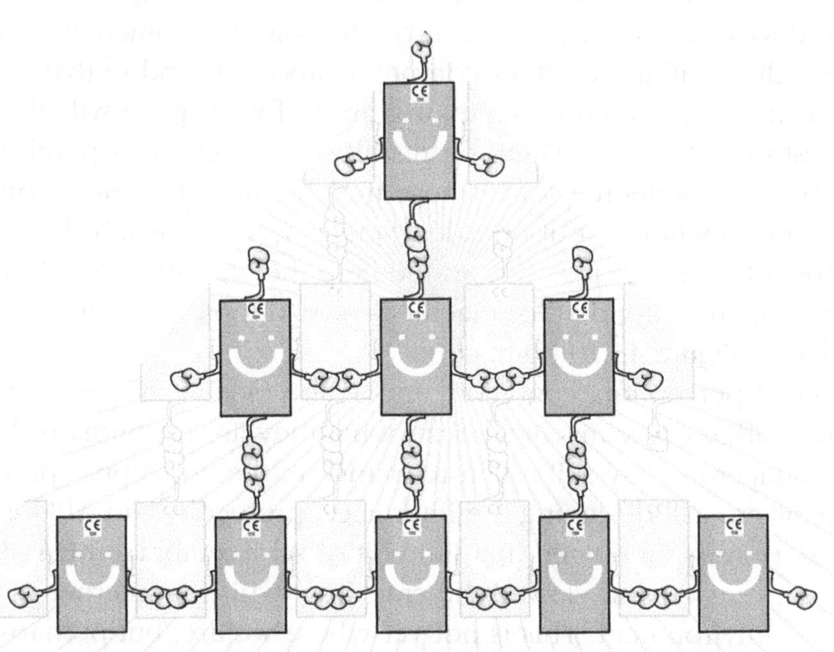

Figure 11.9 What if everyone wanted to be a doctor?

will be key here all the while the user wants to take their fate into their own hands and may be severely biased towards finding even a placebo cure. In a way this is like Charles Bronson becoming a vigilante on the streets of New York in the film "Death Wish", but now thankfully with a view towards alleviating and curing diseases. The question here is how can we promote safe and effective use of medical technology in cases where consumers gain access to the technology under the desk or apply it so far off label that the label might as well be written in the Elvish languages of Middle-Earth? Usability engineering will help manufacturers in addressing new groups of users, but it may be less evident how it can also help them address the perhaps whimsical uses of their technology or walk the do-it-yourself enthusiasts back into the expected fold. We hope it can help with this too.

11.9 Usability for machines

In Section 1.4, we briefly discussed the stubborn ignorance of the IEC 62366-1 standard when it comes to machines as users. We can hardly fault the standard here as this omission leads to a very practical emphasis on human users, who will for the foreseeable future be the largest group of medical device users. It is, however, predictable that someday soon computers will be talking to other computers about us and in that context working to ensure the interfaces they use are fit for purpose will also be of great interest to us humans. Think of this like search engine optimization on the web: an e-commerce store whose merchandise does not show up in search engines will lose out on sales. Similarly, if our health data is not machine-friendly we may lose out on the help the algorithms could provide. A usability standard intended for machines would look quite different from IEC 62366-1, as Figure 11.10 points out.

The rules of perception, cognition, and action (see Section 3.5) are quite likely a little different for machines than what they are for humans, but usability engineering may still retain relevance in the equation – particularly if we humans are still designing the interfaces, protocols, and APIs those devices use. Here we indulged in a little bit of sci-fi, embraced HAL 9000 as a prospective user, and viewed the crackling, whirring internet as the potential use environment. This is not yet where we are, but perhaps this is at least popcorn for thought for now?

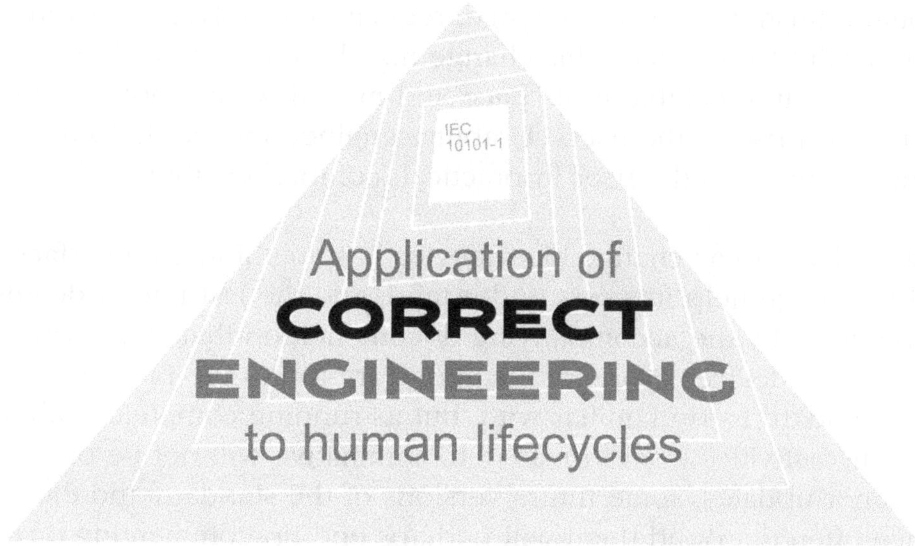

Figure 11.10 What would an UX standard written for machines look like?

Figure 11.11 Evolution is taking us from crawling to walking and to hopefully standing tall.

11.10 Rapid evolution

Getting more serious again, the one universal truth appears to be that change is accelerating all the time. Evolution happens, as Figure 11.11

attempts to illustrate. Devices change, regulations evolve, and even standards want to be agile. This change may be continuous, but it may also be unpredictable in its pace and may, if we are not careful, come as a surprise to the users. Usability engineering can help us introduce changes to the user in practical, comprehensible, and transparent ways.

How and what has changed in a device is something the interface should by design help its users understand, and the last line of defense could be given by the accompanying documentation that ships with updates to the devices. None of the above are topics the present IEC 62366-1 standard is too familiar with, but as running complete usability engineering activities from formative to summative will not be practical for all minor updates, some future versions of the standard and the guidance offered around them will perhaps end up commenting on such matters sooner or later. You and we can, of course, already start considering such factors if the velocity of updates to our devices warrants the concern.

Chapter 12

Conformity Assessment

The number two goal of all standards, right after ensuring organizations are doing the right things, is that other relevant organizations can understand those activities from the outside in. The conformity assessment procedure looks at both of these goals before it can check any box, as depicted in Figure 12.1, *and sign off on the results to any degree.*

Usability is sometimes thought of as both a science and an art. So, how do we rate a painting by Picasso, a concerto by Mozart, or a sculpture by Michelangelo? How do we review the discovery of the DNA structure, for that matter? In conformity assessment of usability engineering activities, we may occasionally be faced with whimsical approaches that reach comparable stratospheres of imaginary flight, but most often the task is much closer to the ground. Instead of starving artistry we have the regulations, standards, and various disciplines of engineering to fall back on. But how do we go about tackling usability? And how do we know we are doing a good enough job addressing the usability of our medical devices?

The IEC 62366-1 standard on usability engineering has an ingrained habit of saying that compliance will be checked by inspection of the usability engineering file. This is all well and good, but what does it actually mean? Afterall the statement is about the same as the first rule of quality management: if something is not documented it didn't happen. The intention is that all the requirements will have to be met in practice and documented in some record matching the activity. But what does this actually mean when assessing the conformity of usability engineering

DOI: 10.4324/9781003645702-12

Figure 12.1 Usability is only really ever about ticking a box in the context of a conformity assessment.

activities and that magical 'usability engineering file' in particular? That is what we'll be discussing in this section.

12.1 Context of the assessment

Conformity assessment of usability engineering activities is most often done by an external party in the context of wishing to place a product on the market through some form of regulatory approval or device registration. The objective of such a checkup is to assess whether a reasonable expectation exists that a product is safe to use as intended and as marketed. In other words, the point is to obtain evidence that any usability-related risks have been mitigated to an acceptable level (e.g., EU 2017/745 Annex I GSPR 5, IEC 60601-1-6, IEC 62366-1, and ISO 14971).

The external party conducting the assessment may be a certification body, a notified body, or a competent authority. The impetus for performing the assessment may even come from the occurrence of some adverse event, in which case the party conducting the checkup activity is usually the regulatory authority of the country where the event in question occurred.

Whether the nudge to assess the activities performed by the manufacturer comes from something having hit the fan or the manufacturer just wishing to place a product on the market, both the process instructions and the product-specific documentation will be of interest to the assessment team. The process in place for usability engineering may occasionally be called 'conformity assessment', 'standard compliance evaluation', or something else by the manufacturer, but a process is nonetheless expected by the standard – and one integrated with the rest of their quality management processes. So, let's begin our conversation here with the standard in the next section.

12.2 Expectations from the standard

The IEC 62366-1 standard begins with the planning of usability engineering activities and then proceeds through their realization to the assessment of their outcome, as you might expect from the universal plan-do-check-act-cycle adopted by all major medical-device standards. In doing so, the standard first wants to see a so-called 'use specification' be formulated (see Section 2.3.1) to describe the use we wish to put our device (or at least its interface) to. This is closely followed by the identification of so called 'safety-related user interface characteristics' (Section 2.3.2). In other words, these first steps focus on understanding the user interface we are working with, its safety and any associated use errors. The activities here might involve little more than pen and paper, as advanced modeling and special tools probably make increasing sense after a first round of intuition has had a chance to have been applied. Depending on the nature of your device, the amount of data you have around its use, and the novelty of your value proposition you may feel differently, of course, but never let the tools drown out common sense – especially at the beginning of asking yourself what should happen and what could go wrong.

The standard then moves the magnification glass to harms, hazards, and hazardous situations (Section 2.3.3), out of which we want to identify any use scenarios related to hazards (Section 2.3.4). Finally, the entire set of such scenarios or some pertinent subsection of them is selected for a summative evaluation (Section 2.3.5). This ultimately leads to a passed summative evaluation, the report of which summarizes our usability activities and their outcome for all interested parties to read. If it happens that a summative evaluation ends in tears it will be followed by appropriate actions and a new summative evaluation.

It is exceedingly rare for all usability activities to go as scripted, at least if the activities have been done correctly, and not just as a paint-by-numbers exercise. In practice some element of iteration usually take place, which is why the standard strongly recommends performing formative evaluations before climbing the summit to conquer the summative evaluation. Identified use errors and hazard-related use scenarios often call for some revision of the use specification, and possibly the intended use or instructions for use. Each iteration may move our camp closer to the summit, but the summit is not reached by visiting any one camp multiple times – in other words, the activity is only done once the expectations on freedom from unacceptable harm are sufficiently met. The standard words this as the usability of the medical-device interface being acceptable from a safety perspective. Any objective evidence that emerges to the contrary can, of course, not be disregarded just because we already gnawed our way to the top before that evidence emerged. To belabor the mountaineering metaphor, the film "Everest" (USA, 2015) based on a true story showed us what might happen when summitting was all that appeared to matter. In the world of medical devices, once the risk (considering both probability and severity) of use errors is demonstrably low enough, and meets the criteria we may have had, we may feel satisfied with the outcome.

One last caution to make here before we dive into looking at the expectations the IEC 62366-1 standard has for your usability engineering process and documentation, is that also other standards will feature in your deliberations. If, for example, your device falls into the category of medical electrical equipment, also specific standards of the IEC 60601-1 family of standards will bring up usability considerations you will need to address.

12.3 Evaluation

Evaluation of conformity to IEC 62366-1 looks at both the process for usability engineering the organization has in place and the documentation this process instructs to be developed. Both of these are then measured again the expectations of the standard and the assessor's understanding of the user and use of the devices developed through the process. Experience with the appropriate healthcare setting (incl. its place in the care flows and any special characteristics it may have), as well as the understanding of any novel operating principles and technologies employed by the devices, are a key factor in conducting a successful evaluation. In the following we'll discuss the process and the documentation separately.

12.3.1 The process

The process for usability engineering is the root of all conformity assessments as it sets up the organization's understanding of usability-related requirements and the way the organization sees itself meeting those expectations. The following list represents high-level goals we expect the process to check in addition to the more detailed requirements the standard sets out for each individual activity. Thus, the process for usability engineering:

- Integrates into the overall QMS of the organization
- Observes all applicable regulatory requirements based on the jurisdictions the manufacturer's products ship to
- Facilitates good communication between the relevant departments and activities involved in the development, release, and maintenance of the medical device
- Ensures that usability engineering activities are complete and judged satisfactory before any product ships
- Ensures that all documentation is satisfactorily complete

The standard then gives us the marks to hit, just like taped crosses on the floor might designate the spots to stand at for an actor delivering a towering monologue at the end of some court room drama. The place is carefully measured for optimal lighting and lens focus, but the point itself does not give the actor the words to utter. In usability engineering, too, the clauses of the standard give the organization the points to consider. The process they then sketch around those markers is entirely up to them, and it should translate the requirements into practice.

IEC 62366-1 expects usability engineering to slot into the tailored overall ISO 13485 quality management system the organization runs. This must link usability engineering with risk management and product development activities in a way that leads to:

- Risk analysis that covers use errors, hazards, and hazardous situations (and vice versa: all usability-related harms are addressed via ISO 14971 risk management)
- The connection between risk management and task analysis, if used, is made in a meaningful way
- The connection between risk control activities and the specification of use is made in a meaningful way
- Appropriate usability considerations being applied throughout design and development activities (including ISO 14971 inherent safety and protective measures)
- Appropriate usability considerations being applied in the creation of accompanying documentation, warnings, and safety symbols (including ISO 14971 information for safety)
- Safe medical devices being manufactured

The usability engineering process must also work for those expected to follow it. This is naturally true for any quality management process, but the usability activities are often performed by a separate group of specialists, or performed as a separate singular activity, both of which may lead to breakdowns in communication between the activities and the teams. It may also be that, for example, the software developers working on the code behind the user interface and the domain specialists analyzing the use of the interface don't naturally talk the same design language or fully appreciate each other's points of view. The process should thus ensure that the process owners and developers hear and understand each other so that the relevant information is exchanged, and the outcomes and implications are understood on both sides. Ultimately, it's perhaps the developers who need to understand what the usability engineering activities mean, and then run with the results to see all issues appropriately resolved in the device. At the same time, it is only practical for the usability engineering specialists to know what they ask for in remedying an issue: asking for something that the developers feel is impossible will perhaps lead to a dismissal of the issue even when some other simpler solution could have solved the issue to everyone's content. Neither activity is intrinsically more important than the other, of course, but as design and development will probably greatly outlast usability engineering, every effort should be made to ensure the latter is understood from the point of view of the former.

Every organization has its own take on meeting and living with requirements of ISO 13485. The design and development processes will differ and the role of usability engineering in these will also differ. The nature of the product (including complexity, risks) will greatly impact any overall web of processes, but so too will the size of the company, the suppliers and consultants frequently contracted. The conformity assessment will listen to all such particulars and then measure compliance against the whole IEC 62366-1 standard.

Typical flaws encountered in the usability engineering process include the following:

- The process is not connected with the rest of the ISO 13485 quality management processes (e.g., design and development, risk management).
- The process is instructed vaguely on an abstract level that does not tie activities into the practical arrangements of the organization (this will lead to reinventing the wheel every time it is needed, and may thus cause resources to be wasted and errors to be made).
- The process does not define clear roles or responsibilities within the organization, which will also lead to issues in demonstrating personnel competence.
- The interplay between usability engineering, risk management, and design and development activities is not understood, or usability engineering is seen as a separate, once-off activity during product development.
- No expectations, requirements, or acceptance criteria are set for usability engineering activities beyond that they are just done.
- The process does not acknowledge the importance of transport, storage, installation, operation, maintenance and repair, or disposal activities as part of the usability of a product (these are fairly commonly forgotten during usability assessments).
- The process does not adequately explain the documentation expected out of the usability engineering activities.

After the initial release of a product, the role of the post-market surveillance (or post-production information) may increase. This may be particularly true for looking at the often-neglected usability issues encountered during transport, storage, installation, operation, maintenance and repair, or disposal of a medical device. The availability and quality of such information will become a topic of interest as theproduct ages on the market.

12.3.2 The documentation

The next component to the conformity assessment is the usability documentation which is often referred to as just the usability engineering file – although the process discussed above could also be seen as a part of this file. In ISO 13485 parlance the process is a document (one or more) and the documentation we are now about to discuss are records (see Juuso 2022 for more discussion on documents and records). The following list represents high-level goals we expect the body of documentation to hit in addition to the more detailed requirements the standard sets out for each individual activity. Thus, the body of developed documentation:

- Matches with the expectations set by the process and by the standard
- Provides a convincing analysis of the usability of the medical device made with the benefit of appropriate competences and resources
- Provides a definitive verdict on the completeness and acceptability of the usability evaluation made by the appropriate authorities at the organization
- Provides adequate information needed for any further development activities pertaining to the device or its documentation
- Provides adequate information (including specifications) for the safe use of the device

As the standard itself points out several times throughout its clauses, conformity with the standard is assessed through usability documentation. This means that no conformity assessment can take place without an assessment of the usability documentation created. This documentation is often referred to as the usability engineering file, which may be a single file or a network of documents and records under the organization's management.

After assessing the process itself (see Section 12.3.1), attention turns to all the documentation instructed by the process and expected by the standard. In addition to the usability engineering standard (IEC 62366-1), also the standards on quality management (ISO 13485), risk management (ISO 14971), and IEC 60601-1-6 come into play here. Any other related standards claimed by the organization may also be of interest, as naturally will any relevant applicable regulatory requirements.

In practice, the body of documentation expected consists of the following types of records:

- Usability engineering plans
- Detailed usability requirements
- Usability criteria which may be defined in plans or separately
- Usability engineering reports
- Product-specific report demonstrating conformity with the set requirements
- Accompanying documentation (including instructions for use)
- Warnings and labeling (usually as part of accompanying documentation)

Conformity assessment is, in essence, conducted by pulling out the IEC 62366-1 standard and marking each individual requirement stated by the standard as either passed or failed based on the evidence provided by the documentation. Each clause and subclause of the standard may make multiple requirements, so that each individual requirement is passed or failed, and not just the top-level clause above them. It is in the manufacturer's best interest to ensure the answers to these requirements can be easily identified in the documentation as this may speed up the assessment process and avoid unnecessary back-and-forth conversation between the assessment team and the manufacturer.

A pertinent example is given by IEC 62366-1 Clause 4.1.3 on information for safety. The clause requires that when information for safety is used as a risk control measure, it is subjected to the usability engineering process to determine that such information is a) perceivable, b) understandable, and c) supports the correct use of the device. During conformity assessment we would thus check that:

- The process acknowledges that information for safety as a risk control is subjected to usability engineering activities.
- The records specify whether information for safety has been used as a risk control measure.
- The records show that such information for safety has been addressed in usability engineering activities.
- The records provide evidence that such information for safety has been found perceivable, understandable, and to support the correct use of a device.

The above expectations are often easily overlooked during development but at the same time quite straightforward to check during an assessment. Your process and documentation templates should thus ensure this requirement is addressed by your usability engineering activities.

Assessing each clause and requirement of the standard thus yields a fairly long test report that essentially corresponds to a pass/fail-annotated version of the standard. Unlike in the case of most regulatory audits, it is possible to discuss the assessment both in advance and after its completion with the certification body. We recommend you take advantage of this opportunity to observe where your solutions are exemplary and where they may fall short, or only just clear the bar. Clearing all the requirements is more like running a marathon than a playing a round of Quidditch with Harry Potter – hard work instead of sorcery is the name of the game here. Instead of the Golden Snitch dreamt up by J.K. Rowling, the occasionally equally elusive documented evidence is what is on everyone's minds during the assessment.

The final test report will reflect all the passes and fails observed when comparing the manufacturer's documentation with that expected by the standard. After receiving the report, the manufacturer should address the observations, and once these are all fixed, perhaps try again. Thus, the report serves as a helpful status check along the path to conformity. If, however, the conformity assessment was performed with the objective of obtaining an accredited certificate on the standard under the certification body scheme (CB), even a single "Fail" will prevent the certificate from being written.

Typical flaws encountered in the usability engineering documentation include the following:

- **Vague or missing requirements**
  Plans, acceptance criteria, and any definition of done are often loosely worded and written in an abstract manner so that it becomes problematic to grasp what the manufacturer required of themselves in the first place – an outside auditor is thus left to wonder whether the performed activities and recorded evidence even convinced the manufacturer themselves.
- **Forgotten information for safety**
  Information for safety is used as a risk control, but it is not addressed in the usability assessments – it is thus left in the air whether the risk

control supports the safe use of the device, or not. A minimal approach here might be that a medical expert reviews the information as appropriate for the care context, while at the other end of the spectrum a whole series of usability tests involving a range of scenarios might be performed.

- **One document for the archives**
  A single ever-growing document attempts to address all of the usability-related questions surrounding a product – it thus becomes questionable whether the usability engineering process integrates into the overall QMS or device lifecycle. Usability engineering becomes a phase in the story of the QMS and the product but does not evolve with these. Technically it's not wrong to put everything in a single document, but it may not be all that plausible to an external auditor that it is then also maintained.
- **Terminology is inconsistent**
  If the terminology used by the manufacturer is not consistent across their documentation or doesn't meet the expectations of the standard, it may be a sign that the manufacturer's process is immature. This raises red flags on how well the expectations are addressed. It may then be necessary to spend time on explaining and mapping the terminology used.
- **Changes are controlled insufficiently**
  It is not enough to build something once and then leave it be. The way the device is used and how it should be used will both evolve over time, and the interface of the device will also be revised occasionally. All such changes will affect the usability of the device, and these changes should show up in our records.

The above list shows only a few of the issues we might expect to see during conformity assessments. This book contains many more in its other sections, but these are perhaps the most frequent issues we have come across. If you compare the above lists of flaws detected in the process and those observed in the documentation you will see that they share much of the same DNA. All of the flaws go back to key expectations of the standard put forward in the earlier sections of this book. The process and the documentation should form a coherent set of information on activities undertaken and the results obtained should speak the same language to the highest extent possible.

12.3.3 The UOUP-path

As discussed in Section 2.5, an abbreviated path exists to conformity with the standard. This path is intended for so-called user interfaces of unknown provenance (UOUP), which may not meet all of the present requirements of the standard due to having grown up without knowing about the standard. The full path essentially looks at the entirety of the standard, mainly Clauses 4.x and 5.1-5.9. The UOUP path is instructed in the normative Annex C of the standard.

Having a simplified path to conformity is naturally a tempting proposition, but it may not always be the right path to choose. The present IEC 62366-1 standard came out in 2015 and was amended in 2020. It is thus increasingly difficult to argue that an interface was developed "a long time ago" without the benefit of the standard. If a product has been on the market for a long time without any changes having been made, now some review and perhaps revision might be even more called for.

Where the UOUP path may, however, prove unexpectedly useful is in the case of a complex interface once developed according to the standard's expectations but now somehow has fallen behind the present (or future) version of the standard. Such an interface was once evaluated to meet the expectations, but now following the UOUP path may allow it to build a simpler, smaller bridge forward. An example here might be a legacy interface that is now scheduled to get a limited upgrade in some limited area, but the rest of the interface remains the same. In this case the unchanged majority of the interface might be addressed as UOUP and only the more limited section of the interface subjected to the full standard. This approach might be supported by the complexity of the interface, the vast body of evidence on its conformity back in the day (e.g., formative assessments, summative assessments, documentation) all of which may also have been addressed via IEC 62366-1 conformity assessment test reports.

In practice, attempting to stick with UOUP over any extended periods of time is not conducive to efficiency or peace of mind. Once lines are drawn between legacy and the more modern areas of the interface, and once changes are enacted over time, the whole may become an entangled mess. The cold war saw interesting solutions being worked out in Berlin which was split between the east and the west – including a metro line running past walled-off enemy stations and over forbidden zones to skip to another part of friendly territory – but that is not a good design choice for medical devices. Our advice here is to discuss your options with your certification body. It will become increasingly laborious to deal with interfaces split into modern and antiquated sections, especially as you may need to make changes in any of them at any point in time.

Typical flaws encountered in UOUP approach to usability engineering include the following:

- The manufacturer's post-production processes fail to observe usability-related information in received feedback information – the resulting analysis is thus missing or inadequate. Subsequently, the UOUP concept can't be used.
- Maintaining risk management activities and information has been neglected – the analysis of the current status of risks becomes impossible and torpedoes any use of the UOUP concept.
- Risk management activities have been maintained, but the changes made to the product or its accompanying documentation are not adequately observed in revisiting activities discussed in IEC 62366-1 Clauses 5.1–5.9 (see Section 2).

For more discussion of the UOUP path, please see Section 2.5.

12.4 The verdict

Over the past 15 years we have been delighted to note that manufacturers' awareness of usability engineering, both its goals and how it should be conducted in practice, have significantly increased. This has had a visible impact on how the usability engineering process has become more streamlined and better integrated into the overall QMS that a manufacturer has. Most importantly, the increased understanding has led to better, safer devices being released although the bad ones do sneak through still occasionally and make headlines across the world.

The poor usability of medical devices thus continues to cause issues in healthcare, home care, and the other contexts medical devices get used in. A device once put through the usability engineering pipeline adequately, but then neglected may over time become equally troublesome in actual use as something where usability engineering was not considered. The need to monitor devices and their use over the post-market period is thus only increasing in importance as many regulations point out.

The appropriate conduction of usability engineering activities calls for not only appreciation of the requirements, expectations, and good practices put forward by the standard and the associated guidance documents, but also the mastering of necessary competences and a diverse enough pool of expertise. In addition to the process and documentation also – and perhaps most importantly – the right people are needed to run the activities, take part in the activities, and pull it all together in a comprehensive yet concise usability engineering file.

Similarly, keeping up with how the device is performing in use, how the standards of care are evolving, and how also the expectations of the user are thus changing in the field is an important part to the maintenance of usability. It often won't be possible to assess these factors in an initial assessment, but as the product gains more laps around the market, the topic will increasingly come up during conformity assessments of usability activities. In computer engineering terms, it won't be possible to spawn new threads of activity for each new device and have these run until perpetuity. It thus becomes even more important to tie usability engineering activities into the overall QMS activities and utilize, for example, post-market surveillance and ECO/ECR change management activities for products to also consider usability aspects. The only certainty is change, and we should plan our conformity activities with that realization to keep everything running along as smoothly as possible. Speaking of that "ever after", let's look at what that may entail in the next chapter.

Chapter 13

Business as Usual

It's time to turn our gaze back to the future, as suggested by Figure 13.1 *and consider how to maintain usability engineering activities over the long term to make the most out of them. Usability may be a factor in the changes we make to the product over time, and it may also help us investigate issues that pop up as the use of the device evolves or we just accumulate more miles in the odometer. Interaction with the user over the long term will be increasingly important, for keeping users in the loop concerning any changes we make, but also as a way of keeping tabs on what shifts in the environment we should be paying attention to through our post-market activities.*

After the discussion in the previous sections of this book, we now have the most usable and useful process for usability engineering defined in our QMS and training to all the relevant staff. We have released medical-device products out there, the usability of which is demonstrably out of this world. What should we do now? This section looks at the lessons to keep in mind over the long term, and how we can keep working towards the elusive level of perfection we may have set our sights on after drinking all the Kool-Aid served to us by the standards. Perfection is, of course, a good aspirational goal, but never a realistic destination for any human endeavor. Instead, here we'll look at some practical lessons we were perhaps too timid to impart in Section 1.

DOI: 10.4324/9781003645702-13

Figure 13.1 Nothing stays up to date by itself – except this simple clock perhaps.

13.1 Focus on safety, branch beyond as resources allow

The IEC 62366-1 standard and the regulations around usability are written from the point of view of safe usability (see Section 1.1), not from any notion of coolness, aesthetic pleasure, pumped-up user experience, or even guaranteed user benefit. In the crossfire of the everyday as changes are suggested, new products and product lines are prepared, user surveys are conducted, and the field in general evolves, our tip is to first focus on safety and then branch beyond this as resources allow. A great user experience and a safe user experience are at least good neighbors, but even if you occasionally get inspiration from the other side of the fence, it's a specific yard you have to keep in check. Similarly, ensuring that the benefits sought are realized over time is a noble goal, but not an excuse to lose sight of how changes are made to your device. Remember that instead of all the perhaps smiling faces you might see in your user base it's actually the fearful and unhappy faces you should approach with more questions.

Ideally you would run usability testing in each and every setting around the world where your device is to be used, and test each task, variation, and permutation until all possible issues are discovered. But as this is never going to be feasible you should focus on safety and work hard to cover this domain. Consider, for example, a story of strange deaths in a single rural hospital always occurring around the same time of day. At first this may sound like a curse or wrath of God, but it turned out to be the work of an untrained janitor removing an electricity plug from the wall to make room for his floor cleaner. A less fantastical example here is given by a visible discharge of static electricity between a metallic hospital bed and a nearby chair that killed a patient monitor and gave a painful shock to the patient prompting him to rip out the monitoring cables from his chest. This may be an issue in low-humidity environments, but not necessarily in environments with above 15% humidity – this too emphasises the need to understand use environments beyond our own individual contexts. Not all of the random acts of variation in life can be accounted for in advance, and not all mitigations are the responsibility of the manufacturers (see Section 1.8), but many factors are, and safety is a good framing for our immediate considerations here.

13.2 Linking usability engineering and risk management

Earlier in Section 9.2 we discussed at length how usability and risks are intwined, and how they can be linked as part of an effective quality management system. This link between the two realms is in many ways the most crucial aspect of usability engineering to get right – so that the right information is generated in the right places at the right time, and this information is passed on and reacted on in the right ways. Recreating similar looking mechanisms in both usability engineering and risk management, or duplicating effort between the two domains, is most likely not going to be effective. Both processes have their unique needs and will evolve over time on their own, but ensuring a connection between the two is maintained will be key to maintaining usability over the long term – over any changes to the product, the QMS, or the many standards and regulations that affect the overall system.

In other words, there is a lot to say on the nuts and bolts of usability engineering, but perhaps the most important connection to pay attention to in the long run is the connection between usability engineering and risk management, and vice versa. Building a solid bridge across this divide may be the quaint little bridge Winnie the Pooh plays Pooh sticks on, or it may be something akin to the bridge across the river Kwai, but we feel we can't stress enough how important this bridge is for sustained usability activities. This connection is the red telephone line between the two superpowers of medical-device development. New issues will be discovered at both ends of the line, and the other end will often need to be consulted or kept in the loop when this happens. Ensuring this line goes in both directions and picking up the phone isn't too laborious or scary is, for us, an everyday magic trick to practice until it becomes second nature.

13.3 Sustained usability

While on the topic of sustained usability, let's go a step further. Having an immaculate track record for a device that has been on the market for years is self-evidently a merit. Still, a device that has been patched several times and has received new features time and time again may be prone to breaking at one of the many seams. Controlled design and development activities, as well as risk management and post-market activities, should help us notice such cracks ahead of time or as soon as possible, but don't just think that something once made will stay in shape as long as it is not touched. Especially if your risk management activities have consistently given perfunctory reports of no new risks being detected, there may be a good reason to revisit usability engineering activities. The choice of when and how such activities are rerun is up to you, but the excuse to dive deeper may also be given by obtaining evidence that your intended benefits have also made an appearance in the wild.

There are enough documents and activities in quality management running with expiration timers already to not make usability into one more such activity. Instead, usability should be a factor in what you already want to do: how you design and develop products, how you perform post-market surveillance, how you conduct corrective and preventive actions around a product (or a process), and how you monitor and analyze risks. Creating a separate thread for ensuring the usability of your product is maintained appropriately may make sense if you have the resources for it, but for smaller companies we would urge thinking about usability as more of a point of view than a process the standard expects. That's a heretic notion, yes, but we'd rather see usability engineering alive and embedded into the activities of the organization than as something only momentarily used before putting out to see.

13.4 Bad usability is more newsworthy than good

We often joke with the readers we meet that if they have positive feedback on our books or courses, they could consider writing reviews somewhere online, but if they have something negative to say they could perhaps discretely e-mail us instead. Similarly, customers of any online service are more likely to write reviews fueled by negative experiences than positive ones. The one exception to this is perhaps a review of a trip made through some ridesharing service, where just about any ride you can walk away from is a five-star ride.

In medical devices, a good user experience is no doubt a selling point. A user impressed by the usability of a device may report on their experiences to a colleague, and they will, of course, be unlikely to sue you as the manufacturer. Yet almost all of the usability-related medical-device stories that gain publicity are negative. Perhaps the most infamous of these is the Therac-25 case from the late 1980s. This was a computer-controlled radiation therapy device, which due to usability-related software errors sometimes gave patients 100x doses of radiation harming or even killing them. Keep this in mind when arguing why some usability engineering activities should be performed. Not all devices will be capable of such dire misfires, but the effects may be negative and they will always include at least potential damage to your brand.

In other words, getting fancy trophies on good usability would of course be great, but not generating body bags is reason enough to rally behind usability engineering. No news may be good news for the usability of a medical device.

13.5 Evolution of the intended use

The origin of every medical device is its intended use, or if that isn't the seed it grew out of, the intended use is at the very least the nexus of the device and the regulations that apply to it. The intended use is the statement of use for the device which anchors it firmly into the domain of regulatory expectations: what the manufacturer promises their device does and what it is good for accomplishing.

In this sense, Section 1.1 of this book should have been on the intended use, and had we written a book about medical devices instead of usability engineering for medical devices it would have been just that. Here at the end of the book, after going through the entire usability engineering process, we bring up intended use again with the sole purpose of acknowledging that you may in fact be thinking about how to expand or improve the intended use over time.

It is perhaps to be expected that many new manufacturers wish to coin a broad enough intended use at the start of their journey so that they will never have to consider a change to that description and yet have a chance to evolve their device in any way they may choose. The footnote often overlooked in penning such loose definitions is that a broad intended use will also require equally broad scientific proof, usability evaluations, and clinical evidence. If you try to plant your flags far and wide you better have an army of horses and personnel to secure the borders of those vast holdings – or perhaps scale your foray more appropriately.

Now at the end of the book is the time we might consider branching out of the box we have painted for ourselves or solidifying the evidence we have to back up some of the key claims. The metro map we presented in Section 10, particularly its track on the post-market clinical follow-up, should have you covered for shoring up your clinical claims on either front, but if it is evidence on the usability of the device that you seek, then don't overlook the tools IEC 62366-1 offers. The earlier sections of this book have discussed, for example, formative and summative usability activities at length and we wish to point out that you can use both to move ahead also after releasing a product. Summative usability evaluations have a hallowed standing, but also formative activities will help you move in the same direction. While not a substitute for clinical evidence, both will help you work on the desired intended uses of your devices.

13.6 A million questions on conformity

At the start of the journey to conformity with usability expectations it may feel like we need to observe a million different things and perhaps prefabricate answers on at least as many questions regarding conformity so that we can pull these answers out during audits. It is only natural to feel like that at the start when the standards, regulations, and possibly even the product itself feel vague and unknown to us. Throughout the previous sections of this book, we have, however, tackled each and every aspect of usability the IEC 62366-1 standard considers essential to ensuring a level of safe usability for medical devices. In going through the sections, we have seen the essential characteristics of both the usability engineering process and the device-specific usability engineering files expected of us. All of this has helped us focus on what matters the most and to shape our processes and the set of documentation around that core of expectations.

After all that work, you will now have a great understanding of the skeleton requirements around which everything in usability engineering revolves. Even if you want to position the user as the center of that universe, the framework of expectations has allowed us to see the exoskeleton of addressing the user, use environment, and use of our device. Based on this knowledge, you will now be in possession of the key answers to the types of questions others may pose to you. You no longer need to generate a phonebook's worth of prefabricated answers to every possible answer just so you feel confident explaining your approach to usability whenever an employee, manager, business partner, customer, or auditor pops up to ask questions. You have now seen the 'Matrix' behind usability, so just like Neo in that film use the knowledge to sort all the incoming bullets on to safe landing trajectories. More than that, use the gained wisdom to fashion better, safer, and more user-friendly medical devices. With that in mind, the first person to always convince on any question is yourself – after you are convinced it will be that much easier to convince others.

13.7 The last hard truths

The last practical piece of advice we want to press here is this: get out there. Talk to your users in whatever capacity you can. Seek them out at conferences and trade shows. Call them up. Send surveys and even Christmas cards. Ask them how they are doing and what irritates them in using your device, in the work they do, in life in general. And most of all: listen carefully without cutting them off, only entertaining them with your stories, or blasting them with your commercial messages of the day. You never know what you learn, and sometimes it takes a little time for the message to sink in or resurface during a later product change you consider. As hard truths go, this was actually a soft one.

Chapter 14

Conclusions

Ostriches don't really bury their heads in the sand. That is just a deeply ingrained, popular misconception and a figure of speech we invoked here in the preface of this book. In reality, as depicted in Figure 14.1, *it only looks as though ostriches drive their heads in the sand when they are tending to their eggs in the nest – in other words, actually looking after the future. Tending to the usability of medical devices, too, is a longer commitment than just a one-time checkbox activity. How frequently and with what triggers you invoke new usability activities or how you employ usability engineering to inform your post-market and change control processes will depend on your product. What matters most here is that you don't think that something done once will stay in great shape from there on out.*

Given everything we have discussed between these covers now, what is usability? Phenomenal usability might be a medical device you'll intuitively use in a safe way as soon as you get it in your hands. Great usability is a medical device you won't have any trouble in using safely once you're first told how to use it. Good usability is a medical device you'll be able to use safely with the help of some accompanying instructions or a little head scratching. Mediocre usability is a medical device that mostly does what you want it to do, and horrible usability is a device that sucker-punches all expectations of what should happen. We can write in more intermediate categories and set pseudo-scientific threshold values for classification, but the above examples set out the grand strokes of the usability space fairly convincingly.

In some ways usability engineering is a self-evident task with a clear-cut goal. We want to understand the user, let the user understand the device,

DOI: 10.4324/9781003645702-14

Figure 14.1 There's more to ostriches than meets the eye.

and tango forward as both the user and the manufacturer expected. In other ways usability engineering is a game of touch football on the roof of a trans-Atlantic jumbo jet flying between the continents. The goal posts are moving constantly, the weather conditions change, bird strikes and meteors may happen, the understanding of what we are doing is evolving constantly – and so too are the risks and expectations of each moving part in the game. A medical device we released when we embarked on the journey may have evolved to almost beyond recognition during the journey, the manufacturer's gaze may have shifted towards a different destination, the user may have figured out more worthwhile goals or applied the device to some completely unanticipated tasks by the time we ever land. Even the rules of air travel may have evolved while we were airborne, and the audience we thought we had onboard may have morphed to something completely different.

The comedian George Carlin famously had strong opinions about the appropriateness of the expression of traveling on a plane instead of in a plane, but regardless of that, touch football is, of course, not a good idea in either context. As an example to bring up here it is an admittedly daft one, but also a somewhat illustrative example. The point of the story was to drive home the fact that even phenomenal usability may morph over time if the expectations around the medical device evolve. It is not possible to create futureproof usability, only to maintain usability on a satisfactory level as the components of the device, its surroundings, and the expectations all the stakeholders (including the manufacturer) have for it inevitably evolve. A 'user interface of unknown provenance' (UOUP, see Section 1.18) might as well be a 'user interface of unknown providence' if we don't plan on keeping tabs on it and maintaining it as necessary. The good news here is that all medical-device standards and regulations call for increasing attention to the long game, the happily ever after. Follow-up is nothing unique to usability engineering and it should not take place in yet another silo, but instead make good use of post-market surveillance activities where it can bring real benefits to the tracking and analysis of risks and shifting expectations.

We hope this book has given you a practical introduction into safety-conscious usability engineering for medical devices, and given you new ideas on how to infuse usability engineering into your own processes and activities. For us the process of going on this dive into usability has been immensely enjoyable and even addictive to the point that we might return with some new medical-device topic in the not-too-distant future – even if this would mean venturing beyond a perfect trilogy and going for a quadrilogy.

Thank you for reading our book! Please send us any feedback you might have, and if the book meets with your approval also pass the word. We wouldn't get to write these books without you picking up a copy, and for that we are truly grateful.

–Authors

References

Johnson, J. (2000) *GUI Bloopers: Don'ts and Do's for Software Developers and Web Designers*. Morgan Kauffmann, San Francisco.

Juuso, I. (2022) *Developing an ISO 13485-Certified Quality Management System – An Implementation Guide for the Medical-Device Industry*. Routledge, New York.

Juuso, I. & Pöyhönen, I. (2024) *Medical-Grade Software Development – How to Build Medical-Device Products That Meet the Requirements of IEC 62304 and ISO 13485*. Routledge, New York.

Juuso, I. & Seppänen, T. (2023) *Service Design for Medical Devices* (a chapter in the book Service Design for Emerging Technologies Product Development – Bridging the Interdisciplinary Knowledge Gap). Springer Publishing, Cham.

Nielsen, J. (1994). *Usability Engineering*. Morgan Kauffmann, San Diego.

Index